SpringerBriefs in Agriculture

SpringerBriefs in Agriculture present concise summaries of cutting-edge research and practical applications across a wide spectrum of topics in agriculture with a fast turnaround time to publication. Featuring compact volumes of 50 to 125 pages, the series covers a range of content from professional to academic. Monographs of new material are considered for the SpringerBriefs in Agriculture series. Typical topics might include: A timely report of state-of-the art analytical techniques, a bridge between new research results, as published in journal articles, and a contextual literature review, a snapshot of a hot or emerging topic, an in-depth case study or technical example, a presentation of core concepts that students must understand in order to make independent contributions. Best practices or protocols to be followed. A series of short case studies/debates highlighting a specific angle.

More information about this series at http://www.springer.com/series/10183

Constansia Musvoto · Karen Nortje ·
Anton Nahman · William Stafford

Green Economy Implementation in the Agriculture Sector

Moving from Theory to Practice

 Springer

Constansia Musvoto
Natural Resources and the Environment Unit
Council for Scientific and Industrial
Research
Pretoria, South Africa

Karen Nortje
Natural Resources and the Environment Unit
Council for Scientific and Industrial
Research
Pretoria, South Africa

Anton Nahman
Natural Resources and the Environment Unit
Council for Scientific and Industrial
Research
Pretoria, South Africa

William Stafford
Natural Resources and the Environment Unit
Council for Scientific and Industrial
Research
Pretoria, South Africa

ISSN 2211-808X   ISSN 2211-8098   (electronic)
SpringerBriefs in Agriculture
ISBN 978-3-030-01808-5   ISBN 978-3-030-01809-2   (eBook)
https://doi.org/10.1007/978-3-030-01809-2

Library of Congress Control Number: 2018964249

This Springer imprint is published by the registered company Springer Nature Switzerland AG
The registered company address is: Gewerbestrasse 11, 6330 Cham, Switzerland

Preface

The green economy is a relatively new (and evolving) concept, which started gaining traction following the 2008 financial crisis, and in the run-up to the 2012 United Nations Conference on Sustainable Development in Rio de Janeiro. The concept has subsequently been adopted by a number of countries as a framework for achieving sustainable development, while at the same time growing their economies. Although the term 'green economy' is often interpreted in a narrow environmental sense, many of the principles of a green economy address socio-economic imperatives and human well-being issues.

Agriculture is a key sector for driving the green economy and is central to a green economy transition in Africa and other developing regions. Currently, there is a large body of knowledge on the theoretical aspects of the green economy. However, the translation of this abstract concept into practice has remained largely unexplored, as evidenced by the paucity of publications on green economy implementation. There is thus a knowledge gap between green economy theory and practice and a need to build the knowledge base required to inform green economy implementation in specific sectors. This book, centred on green economy implementation in the agriculture sector, focusing on vegetable production contributes towards addressing this knowledge gap. It also contributes to the body of knowledge on the development of green economy implementation techniques.

In a sector such as agriculture, operationalizing a fluid, multi-faceted concept such as the green economy is not simple. Green economy implementation entails making sense of diverse social, economic and environmental information sets from various sources—from the general to the specific and from the global to the local and translating that information into systematic actions. Furthermore, implementing the concept is complicated by the fact that agriculture is a primary sector which tends to be resource-intensive and can have adverse impacts on the environment, which is counter to the principles of a green economy.

Informed by desktop reviews and field research in the form of case studies of small-scale vegetable production in South Africa, this book examines both the theoretical and practical basis of green economy implementation, and highlights the key factors that have to be considered when implementing green economy projects (in general and in the agriculture sector in particular), with a specific focus on small scale vegetable production. The book examines the areas of divergence between the green economy and crop production, including ways of addressing the tensions and trade-offs that arise. In addition, the book describes a green economy implementation framework, which includes a methodical process to follow at the project level. The framework is adaptable for green economy implementation in other sectors as well.

The book is primarily targeted at researchers, specialists, and students in the fields of green economy, agriculture, and sustainable development and is intended to contribute to the body of knowledge on the development of green economy implementation techniques. The book also has relevant information for agricultural and green economy practitioners, development agents, policy makers and practitioners in both the public and private sectors.

Chapter 1 provides context for the book by explaining some of the key concepts and terms. Chapter 2 discusses the socio-economic factors that should inform green economy implementation, particularly highlighting these issues in a developing country context. Local, national and global factors are analysed. Biophysical and environmental factors that have implications for green economy implementation in small scale crop production are covered in Chap. 3. Chapter 4 covers field case studies of small scale vegetable production conducted in South Africa. The case studies provide insights into the implementation of green economy projects in a specific setting. Chapter 5 presents a green economy project implementation framework which was developed through a desktop review and the field case studies described in Chap. 4. The framework includes a process for consolidating all the information that should inform implementation of a green economy project into a number of key considerations and a methodical process for executing the considerations. Chapter 6 synthesizes information from Chaps. 1 to 5 and provides recommendations for green economy project implementation.

Pretoria, South Africa   Constansia Musvoto
                                                              Karen Nortje
                                                              Anton Nahman
                                                              William Stafford

Contents

Acronyms

4IR	Fourth Industrial Revolution
AAA	Adaptation of African Agriculture Initiative
ACPC	African Climate Policy Centre
AfDB	African Development Bank
AFRICEGE	African Centre for a Green Economy
AGRA	Alliance for a Green Revolution in Africa
ASFG	African Smallholder Farmers Group
AU	African Union
CA	Conservation Agriculture
CARA	Agricultural Resources Act
CGIAR	Consultative Group on International Agricultural Research
CIAT	Centro Internacional de Agricultura Tropical
CSA	Climate smart agriculture
CSF	Critical Success Factors
DEA	Department of Environmental Affairs
DEAT	Department of Environmental Affairs and Tourism
DEDT	Department of Economic Development and Tourism
DRDL	Department of Rural Development
DTI	Department of Trade and Industry
ECA	United Nations Economic Commission for Africa (also known as UNECA)
EDD	Economic Development Department
FAO	Food and Agriculture Organisation of the United Nations
GAP	Good Agricultural Practices
GDP	Gross Domestic Product
GEO-5	Global Environment Outlook
GGP	Gross Geographic Product
GHG	Greenhouse gas
GIS	Geographic Information Systems
GMOs	Genetically modified organisms

GR	Green Revolution
GTEDA	Greater Tzaneen Economic Development Agency
GTLM	Greater Tzaneen Local Municipality
GVCs	Global Value Chains
IAASTD	International Assessment of Agricultural Knowledge, Science and Technology for Development
IDPs	Integrated Development Plans
IEG	Independent Evaluation Group
IFAD	International Fund for Agricultural Development
ILO	International Labour Organisation
IoT	Internet of Things
IPCC	Intergovernmental Panel on Climate Change
IPM	Integrated Pest Management
ITU	International Telecommunication Union
IWMI	International Water Management Institute
IWRM	Integrated Water Resources Management
LDARD	Limpopo Department of Agriculture and Rural Development
LED	Local Economic Development
MEA	Millennium Ecosystem Assessment
MTSF	Medium Term Strategic Framework
NDP	National Development Plan
NEET	Not in Employment, Education or Training
NEPAD	New Partnership for Africa's Development
NFSD	National Framework for Sustainable Development
NPC	National Planning Commission
NSSD	National Strategy for Sustainable Development
ODA	Official Development Assistance
OECD	Organisation for Economic Cooperation and Development
PAGE	Partnership for Action on Green Economy
RIASCO	Regional Interagency Standing Committee for southern Africa
RSA	Republic of South Africa
SADC	Southern African Development Community
SAGCOT	Southern Agricultural Growth Corridor of Tanzania
SDGs	Sustainable Development Goals
SOM	Soil organic matter
StatsSA	Statistics South Africa
TRALAC	Trade Law Centre
UBPL	Upper-bound poverty line
UN	United Nations
UNCTAD	United Nations Conference on Trade and Development
UNDESA	United Nations Department of Social and Economic Affairs
UNECA	United Nations Economic Commission for Africa
UNEP AI	United Nations Environment Programme Aridity Index
UNEP	United Nations Environment Programme
WEF	World Economic Forum

WFP	World Food Programme
WHL	Woolworths Holdings Limited
WRI	World Resources Institute
WWF	World Wildlife Fund

List of Figures

List of Tables

List of Boxes

Chapter 1
Agriculture in a Green Economy

1.1 What Is a Green Economy? Origins, Definitions and the Link Between a Green Economy and Sustainable Development

The concept of a 'green economy' was first introduced by Pearce et al. (1989); in a book entitled "Blueprint for a Green Economy." However, it was not until the global financial crisis of 2008, which coincided with increasing recognition of deepening global environmental and social crises that the concept of a 'green economy' moved into the mainstream of policy discourse. The United Nations (UN) General Assembly and other UN agencies saw the crisis as an opportunity to incorporate 'green' investments in the stimulus packages being implemented to spur economic recovery (Ocampo et al. 2011). The United Nations Environment Programme (UNEP) called for a 'global green new deal' (Barbier 2009; UNDESA 2009), based on a package comprising public-sector investments as well as policy and pricing reforms, with the aim of sparking a transition to a green economy, while at the same time stimulating economic recovery and addressing unemployment and poverty (UNEP 2011).

The green economy therefore came to be seen as a way of simultaneously reviving the global economy and creating jobs (in both the short and the long term), while alleviating poverty and environmental degradation, thereby contributing to sustainability in the long term (Barbier 2009). "A green economy in the context of sustainable development and poverty eradication" subsequently became one of the two main focal themes at the 2012 United Nations Conference on Sustainable Development (Rio + 20). As summarized by the African Consensus Statement to Rio + 20, "the combined stream of economic, social and environmental crises that have plagued the global economy in recent years points to a need to reorient the current development models towards a more efficient, inclusive and sustainable economy by enhancing the resource efficiency of national economies, and decoupling economic activity from environmental degradation" (African Ministers 2011). The underlying principle is to reduce the dependence of economic development on the consumption of natural

© The Author(s), under exclusive license to Springer Nature Switzerland AG 2018

C. Musvoto et al., *Green Economy Implementation in the Agriculture Sector*,
SpringerBriefs in Agriculture, https://doi.org/10.1007/978-3-030-01809-2_1

resources (i.e. to 'decouple' economic growth from environmental damage), whilst meeting social needs and addressing inequalities.

The green economy concept became popularized by UNEP (2011) in its "Green Economy Report," published in 2011, which set the framework for international discussion, and developed a model for assessing the capacity for 'green investments' to deliver on economic growth and job creation, as well as environmental outcomes. UNEP (2011: 16) defines a green economy as "one that results in improved human well-being and social equity, while significantly reducing environmental risks and ecological scarcities." Specifically, such an economy must be low-carbon, resource efficient and socially inclusive. According to UNEP, the key driver of income and employment in such an economy is public and private investments aimed at reducing carbon emissions and pollution, enhancing energy and resource use efficiency, and preventing biodiversity degradation and the loss of ecosystem services (UNEP 2011). Such investments must be catalysed and supported through targeted government expenditure, as well as policy and regulatory reforms. Natural capital must be maintained, enhanced and, where necessary, rebuilt as a critical asset for generating economic growth and human well-being. This is of particular relevance for poor communities, particularly in developing countries, who rely on functional and intact ecosystems for their livelihoods (UNEP 2011).

At the same time, other large international institutions such as the Organisation for Economic Cooperation and Development (OECD) and the World Bank were developing reports and strategies relating to similar concepts, such as 'Green Growth' (OECD 2011) and 'Inclusive Green Growth' (World Bank 2012). For example, the OECD defines the pursuit of green growth as "fostering economic growth and development while ensuring that natural assets continue to provide the resources and environmental services on which our well-being relies" (OECD 2011: 9). The World Bank (2012: 30) defines Inclusive Green Growth as economic growth that is environmentally sustainable; i.e., that is "efficient, clean, and resilient—efficient in its use of natural resources, clean in that it minimizes pollution and environmental impacts, and resilient in that it accounts for natural hazards and the role of environmental management and natural capital in preventing physical disasters."

Given the multi-faceted nature of the green economy concept, and therefore the various different ways in which the concept can be defined, the term 'green economy' can perhaps best be understood by referring to its principles, which are generally well understood and agreed upon, rather than a specific definition. For example, the United Nations Department of Social and Economic Affairs (UNDESA) (Allen 2012) reviewed eight published lists of green economy principles; and derived a consolidated list of eleven principles based on those occurring with greatest frequency (i.e. five or more times); as follows:

   i.   The green economy is a means for achieving sustainable development.
  ii.   The green economy should create decent work and green jobs.
 iii.   The green economy is resource and energy efficient.
 iv.   The green economy respects planetary boundaries or ecological limits or scarcity.

v. The green economy uses integrated decision making.
vi. The green economy measures progress beyond GDP using appropriate indicators/metrics.
vii. The green economy is equitable, fair and just—between and within countries and between generations.
viii. The green economy protects biodiversity and ecosystems.
ix. The green economy delivers poverty reduction, well-being, livelihoods, social protection and access to essential services.
x. The green economy improves governance and the rule of law. It is inclusive; democratic; participatory; accountable; transparent; and stable.
xi. The green economy internalises externalities.

In this book, we develop a framework in which these broad, aspirational principles can be operationalised and implemented at the project level, with a specific focus on implementation in the agricultural sector (although, the framework can readily be applied to other sectors as well). The following section provides more context regarding the place of agriculture in a green economy.

1.2 The Role of Agriculture in a Green Economy

Agriculture has both direct and indirect effects on people and the environment, and thus has direct linkages to the principles of a green economy and a key role to play in a green economy. The magnitude of the agriculture sector (in terms of land area and use of resources such as water), its impacts on the environment, and its direct role in the well-being of people) places it at the centre of the green economy globally, and particularly in Africa and other developing regions. Globally, agricultural land (defined by the Food and Agriculture Organisation of the United States (FAO) as land used for cultivation of crops and animal husbandry) comprises 37% of the land area; and 42% of the land area in sub Saharan Africa (FAO 2017; World Bank 2018). Crops and livestock account for 70% of all water withdrawals globally, and up to 95% in some developing countries (FAO 2018). Globally, the agriculture sector as a whole provides livelihoods for 2.5 billion people (FAO 2016); while in Sub-Saharan Africa, agriculture is the largest employer, and is also the most immediate means of catalysing economic growth and employment for young people (Filmer and Fox 2014; FAO 2016).

Many agricultural activities could provide solutions to the social, economic and environmental challenges that the green economy aims to address, or worsen them. Agriculture can provide livelihoods and food security[1] for a rapidly expanding population, reduce the risks from climate change, and meet increasing demands for energy in the face of dwindling reserves of fossil fuels (Jensen et al. 2012). Depending on

---

[1] Food security can be defined as a state in which "all people, at all times, have physical, social and economic access to sufficient, safe and nutritious food that meets their dietary needs and food preferences for an active and healthy life" (FAO 2003).

how it is practised, agriculture can directly contribute to meeting many of the green economy's social and environmental aspirations, such as protecting biodiversity and ecosystems, and creating decent work and green jobs. The key role of agriculture in a green economy was highlighted by the FAO (2012a, b) in a concept note prepared for the Rio + 20 conference: "As the single largest sector using 60% of world's ecosystems and providing livelihoods for 40% of today's global population, the food and agriculture sector is critical to greening the economy. There will be no green economy without agriculture". The key role of agriculture in Africa's economic development has also been acknowledged by the African Union (AU). African Heads of State and Government declared 2014 the Year of Agriculture and Food Security in Africa. This high level focus was meant to encourage countries to increase food security, reduce poverty, promote economic growth and create wealth through agricultural improvement (Moodley 2013).

For the agriculture sector to have a place in a green economy, it has to be multifunctional, so that in addition to meeting current and future needs for food and other biomass-based materials, it also needs to provide non-commodity goods and services (such as carbon sequestration), reduce poverty, and enable inclusive economic growth, while not disrupting social and cultural systems. To be in tune with a green economy, agriculture has to meet a number of requirements. These include producing food and agricultural goods and services on a sustainable basis; reducing pollution and using resources more efficiently; and maintaining and increasing farm productivity and profitability (Farming First Coalition 2013a). In a green economy, the agriculture sector has to find ways of meeting the demand for increased output that are more efficient in their use of resources, and less damaging to the environment. In addition, agriculture needs to create sustainable livelihoods[2] for farmers and others along the supply chain (Farming First Coalition 2013b). The following section highlights the specific characteristics of agriculture that make it particularly relevant to a green economy.

1.3 Attributes of Agriculture that Make It Relevant to a Green Economy

The rural nature of agriculture places it in the vicinity of a large, typically poor population, and this is especially the case in Africa; making agriculture particularly relevant for addressing social aspirations of a green economy such as poverty reduction and livelihood provision. The UN Food and Agriculture Organization (FAO 2015) notes that Sub-Saharan Africa's population is expected to remain primarily rural up to 2033, and that the absolute number of people living in rural areas will continue to increase up to 2050. According to Beegle et al. (2016) rural areas in Africa

---

[2] A livelihood is sustainable when it can cope with and recover from stresses and shocks, maintain or enhance its capabilities and assets, while not undermining the natural resource base (Chambers and Conway 1992).

remain much poorer than urban areas. The World Bank (2016) notes that 80% of the worldwide poor live in rural areas, 64% (of the worldwide poor) work in agriculture.

Provided that it is practised in ways that align with green economy ideals, agriculture presents real opportunities for contributing to poverty reduction and for providing humans with a range of other products that are essential to their well-being (Power 2010). For example, the Millennium Ecosystem Assessment (MEA) (2005) notes that well-being cannot be considered in isolation from the natural environment, as the services provided by its ecosystems directly affect well-being. Agroecosystems also produce a variety of ecosystem services, such as regulation of soil and water quality, carbon sequestration, support for biodiversity and cultural services (Power 2010), and these are also directly linked to human well-being.

The agriculture sector, and in particular economic growth within this sector, is a proven driver of poverty reduction (World Bank 2007). There is evidence of strong linkages between agricultural growth and poverty reduction among smallholder farmers (Djurfeldt 2013). The African Development Bank (AfDB) (2010) notes that when agriculture stimulates growth in Africa, the growth is twice as effective in reducing poverty as growth in other sectors. Agriculture not only provides opportunities for rural poverty reduction, it is also able to contribute to the alleviation of urban poverty, as it offers an alternative to migration to urban slums. The UN notes that raising agricultural production (particularly in food deficit countries), while at the same time improving the livelihoods of smallholder farmers and preserving ecosystems, would contribute to rural development; and thereby slow the trend towards urbanisation, and the attendant stress it places on public services in urban areas (UN 2010).

One of perspectives through which agriculture's role in human well-being in the context of a green economy can be understood is the Sustainable Development Goals (SDGs). As the prime connection between people and the planet, sustainable food and agriculture have great potential to address many challenges, including provision of affordable, nutritious food, strengthening livelihoods and others; thus driving positive change across the SDG agenda (FAO 2018). FAO (2015) notes that the SDGs offer a vision of a fairer, more prosperous, peaceful and sustainable world in which no one is left behind. This vision aligns with green economy ideals; and agriculture in a green economy context is critical for the attainment of this vision. According to the FAO, without rapid progress in reducing and eliminating hunger and malnutrition by 2030, the SDGs cannot be achieved (FAO 2015). Agriculture is central to SDG 2 (zero hunger), and this in turn is linked to several other SDGs, including SDG 3 (good health and well-being), SDG 4 (quality education).

Another attribute of agriculture which aligns it to the green economy is its ability (again, provided that it is practised in accordance with green economy principles) to restore and protect the environment. Agriculture could support green economic growth through techniques and practices which sustain production while enhancing (or at least reducing negative impacts on) the resource base and natural environment. For example, while many current agricultural practices contribute to global greenhouse gas emissions, good management practices can result in an almost carbon-neutral sector, as well as the creation of environmental services and the generation of renewable energy, while also achieving food security. Many of the techniques and practices that could be applied to enable agriculture to be productive and to protect the environment are well documented and widely accessible.

Agriculture's ability to mitigate climate change (through sequestering carbon and reducing emissions) is an attribute which aligns it particularly well to the green economy. The main mitigation options include reduction or prevention of emissions by conserving existing carbon pools in soils or vegetation; and sequestration—enhancing the uptake of carbon in terrestrial reservoirs and reducing CO_2 emissions (Smith et al. 2014). Carbon can be held in soil by planting perennials, while emissions can be reduced through minimising the use of inorganic fertilisers, restoring degraded lands and preventing deforestation (Schaffnit-Chatterjee et al. 2011). Techniques such as cover cropping, reduced tillage, no tillage, agroforestry, and improved cropland management e.g. using improved crop varieties, extending crop rotations (particularly those of perennial crops that allocate more carbon below ground), and avoiding or reducing the use of bare (unplanted) fallow can all be used to achieve mitigation (IPCC 2007: 506).

Through its forward and backward linkages to other sectors, agriculture creates both direct and indirect economic and employment opportunities. Although 65% of Africans rely on agriculture as their primary source of livelihood (ECA 2013), agriculture has the potential to create even more livelihood opportunities, through appropriate policies and investments (e.g. supporting agribusinesses and agro-processing). Agribusiness is labour-intensive in terms of creating jobs and generating value-add, and it strengthens agriculture's forward and backward linkages with other sectors; thus creating employment opportunities and increasing incomes, strengthening food security and alleviating poverty (Lopes 2015; Woldemichael et al. 2017). These attributes make agriculture pertinent to the green economy.

Finally, agriculture's direct role in food production is an attribute which links it directly to green economy principles focused on enhancing delivery of social protection and access to essential services. In addition to its current role, the food production capacity of agriculture has the potential to create future livelihood and economic opportunities in Africa. For instance, the value of Africa's food markets is projected to increase from US$313 billion in 2010 to US$1 trillion in 2030 (World Bank 2013). This creates opportunities for local business development linked to food production, processing, distribution and marketing, which can in turn stimulate further agricultural development.

As highlighted above, agriculture has a number of attributes that make it particularly suited to a green economy. However, these qualities on their own do not ensure achievement of green economy ideals. An enabling environment has to be created, and factors which hinder agricultural production and development have to be addressed, if the full potential of agriculture as a driver of green economic growth is to be realised. UN Environment notes that while green agricultural initiatives can potentially contribute towards reducing poverty (UNEP 2011), this requires policy reforms and investments which provide an enabling environment for the sector to reduce poverty through increased yields and creation of new and more productive green jobs. Diao et al. (2010) note that factors such as rural infrastructure, linkages between agriculture and other sectors of the economy; and investments in agricultural research and development determine the effectiveness of agriculture as a contributor to economic growth.

1.4 Concepts of Relevance to Agriculture in a Green Economy Context

The earth's natural resources, especially soil, water, plant and animal diversity, climate and ecosystem services are critical for agriculture. Historically, agricultural development has tended to concentrate on increasing productivity with little regard for the impacts of agriculture on the environment. There is overwhelming evidence of the negative impacts of modern agriculture on the environment and the subsequent cost in the provisioning of ecosystem services (IAASTD 2015). There are concepts and practices which use the term 'green' in agriculture. Most are designed to address the negative impacts of agriculture on the environment; and generally aim to enhance production while protecting the environment and/or enhancing human well-being. While these concepts align with the ethos of a green economy, they should be distinguished from the green economy; and their relevance to green economy implementation should be understood. These concepts are discussed below:

The Green Revolution (GR)
The GR took place between 1950 and the late 1960s; and was characterised by crop productivity growth which was driven by high-yielding varieties of maize, wheat and rice, in association with increased application of chemical fertilisers, agro-chemicals, irrigation and increased mechanisation. The production of cereal crops tripled during the GR, with only a 30% increase in land area cultivated (Wik et al. 2008). While the GR increased crop yields in areas such as south-east Asia, India and South America, the impact on Africa was small. Despite increasing crop yields considerably, the impact of the GR on food security and poverty was less than expected and the yield improvements were sometimes at the cost of natural resources. Burney et al. (2010) note that the GR had unintended consequences, and these include soil degradation and chemical runoff; and these impacts extended beyond cultivated areas. While not linked to the concept of a green economy in an agriculture context, the GR has lessons for green economy implementation, such as the need for more integrative approaches and environmental protection. Green economy implementation thus has to be cognisant of the limitations of the GR and take care to avoid its shortfalls.

Greening Agriculture
The greening of agriculture is a response to the decline in the quality of the natural resource base associated with conventional agriculture. It refers to the use of farming practices and technologies that (i) simultaneously maintain and increase farm productivity and profitability, while ensuring the provision of food and ecosystem services on a sustainable basis (ii) reduce negative externalities and gradually lead to positive ones; and (iii) rebuild ecological resources (i.e. soil, water, air and biodiversity natural capital assets) by reducing pollution and using resources more efficiently (FAO 2012b; UNEP 2011). The greening of agriculture entails protecting natural resources (soil, water, air, plants and animals) by reducing pollution and using resources more efficiently (UNEP 2011) and aligns agriculture with green economy principles.

Conservation Agriculture

Conservation Agriculture (CA) is one of the responses to the environmental degradation caused by agriculture and is applicable to green economy implementation. CA is a way of producing crops while striving to save resources, achieve high sustained production levels and acceptable profits, and conserve the environment (FAO 2015). CA is based on optimizing yields and profits, in order to achieve a balance of agricultural, economic and environmental benefits (Bockel et al. 2011) and is one of the practices that can be applied to improve alignment of agriculture with green economy principles. The FAO (FAO 2017) highlights that CA is characterised by three inter-linked principles, namely: (i) minimum mechanical soil disturbance; (ii) permanent organic soil cover; and (iii) diversified cropping (rotations in the case of annual crops or plant associations in the case of perennials). Minimal soil disturbance facilitates the retention of soil organic matter, which provides nutrients for crops and stabilises soil structure, making soil less susceptible to degradation (FAO 2010). Permanent soil cover protects soil from the impacts of rain and wind, and helps retain soil moisture and stabilise soil temperature in the surface layers, and creates conditions that are conducive for maintaining biodiversity (FAO 2010). Growing crops in mixtures or rotations (in space or over time) helps to control pests and diseases by breaking their cycles, and can suppress weeds and improve soil structure through the penetration of different root systems (FAO 2010). Burney et al. (2010) note that applying CA practices can contribute to sequestering organic carbon in agricultural systems; building fertility and improving yields in degraded soils, and mitigating agricultural greenhouse gas emissions.

Climate Smart Agriculture

Green economy implementation occurs in a context where climate change is already impacting negatively on agriculture, with smallholder farmers being especially vulnerable to climate-related shocks (FAO 2013). Climate Smart Agriculture (CSA), defined by FAO (2014) as an integrative approach to addressing the challenges of food security and climate change; incorporates both climate change adaptation and mitigation. CSA has three objectives: (i) sustainably increasing agricultural productivity, so as to support equitable increases in farm incomes, food security and development; (ii) adapting and building resilience of agricultural and food security systems to climate change at multiple levels; and (iii) reducing greenhouse gas emissions from agriculture (FAO 2014). CSA calls for better integration of adaptation and mitigation actions in agriculture, so as to capture synergies between them and to support and sustain agricultural development under conditions of climate change (Lipper and Zilberman 2018). CSA is one of the concepts that is relevant to green economy implementation in an agricultural context as it addresses some of the issues that a green economy aims to address.

Sustainable Agriculture

The goal of sustainable agriculture is to produce food and other products using farming methods that are economically profitable and that protect the environment, human health and communities. Sustainable agriculture has social, economic and environmental dimensions. Ikerd (1993) described sustainable agriculture as "capable of maintaining its productivity and usefulness to society indefinitely; and such an agriculture must use farming systems that conserve resources, protect the environment, produce efficiently, compete commercially and enhance the quality of life for farmers and society overall". Kirchmann and Thorvaldsson (2000) note that components of agricultural sustainability include maintaining production potential, environmental stewardship, economic viability and social justice. According to WWF (2017), sustainable agriculture is the key to producing food within the capacity of the planet, while maintaining the ecosystem services that agriculture depends upon, like healthy soils, clean water and pollinating insects. Sustainability in agriculture also incorporates concepts of both resilience (the capacity of systems to buffer shocks and stresses) and persistence (the capacity of systems to continue over long periods), and addresses multiple economic, social and environmental outcomes (Pretty et al. 2008). Sustainable agriculture is relevant to green economy implementation at it aligns agriculture with social, economic and environmental green economy principles. Essentially, sustainable agriculture recognises the multi-functionality of agriculture and the interconnectedness of agriculture's different roles and functions in the economic, social and environmental spheres.

1.5 Conclusion

A key first step in successful green economy project implementation is having a good understanding of what a green economy entails and its relevance to a specific sector. Such understanding enables informed application of the concept and its adaptation to a specific project situation. This chapter presents the theoretical basis of a green economy and its relevance to green economy implementation in the context of agriculture. The chapter provides context for the book by presenting information to build understanding of the terms and concepts that are central. It sets the scene for the following chapters by illustrating both the complexity of the term as well as highlighting a number of related terms and concepts that together provide the context within which agricultural projects can be implemented.

References

African Development Bank (2010) Agriculture sector strategy: 2010–2014. https://www.afdb.org/fileadmin/uploads/afdb/Documents/Policy-Documents/Agriculture%20Sector%20Strategy%202010-14.pdf. Accessed 23 Aug 2018

African Ministers (2011) African consensus statement to Rio + 20 following the Africa regional preparatory conference for the united nations conference on sustainable development (Rio + 20), Addis Ababa, Ethiopia, October 2011

Allen C (2012) A guidebook to the green economy. Issue 2: exploring green economy principles. United Nations Department of Economic and Social Affairs (UNDESA): United Nations Division for Sustainable Development

Barbier EB (2009) A global green new deal. United Nations Environment Programme - Economics and Trade Branch, New York

Beegle K, Luc C, Andrew D, Isis G (2016) Poverty in a rising Africa. World Bank, Washington, DC. https://doi.org/10.1596/978-1-4648-0723-7 (License: Creative Commons Attribution CC BY 3.0 IGO)

Bockel L, Tinlot M, Jonsson M (2011). A Multiplication of green concepts in agriculture: building the path towards wide up-scaling. FAO EasyPol issue paper. http://www.fao.org/fileadmin/templates/ex_act/pdf/Policy_briefs/draft_green_concept.pdf. Accessed 03 Oct 2018

Burney JA, Davis SJ, Lobell DB (2010) Greenhouse gas mitigation by agricultural intensification. PNAS 107:12052–12057

Chambers R, Conway G (1992) Sustainable rural livelihoods: practical concepts for the 21st century. IDS discussion paper 296. Brighton: IDS

Diao X, Hazell P, Thurlow J (2010) The role of agriculture in african development. World Dev 38:1375–1383

Djurfeldt A (2013) African re-agrarianization? Accumulation or pro-poor agricultural growth? World Dev 41:217–231

ECA (2013) Rethinking agricultural and rural transformation in Africa. Challenges, opportunities and strategic policy options

Farming First Coalition (2013a) http://www.farmingfirst.org/. Accessed 25 June 2013

Farming First Coalition (2013b) Agriculture for a green economy: improved rural livelihood, reduced footprint, secure food supply. Farming first policy paper on agriculture and the green economy. http://www.farmingfirst.org/wordpress/wp-content/uploads/2011/10/Farming-First-Policy-Paper_Green-Economy.pdf. Accessed 25 June 2013

FAO (2003) Trade reforms and food security: conceptualizing the linkages. http://www.fao.org/docrep/005/y4671e/y4671e00.htm#Contents. Accessed 01 Oct 2018

FAO (2010) Farming for the future: an introduction to conservation agriculture. REOSA technical brief 1. http://www.fao.org/fileadmin/user_upload/emergencies/docs/FAO_REOSA_Technical_Brief_1__July_2010_.pdf. Accessed 27 Sept 2018

FAO (2012a) FAO@Rio + 20: greening the economy with agriculture (GEA) - taking stock of potential, options and prospective challenges. Concept note. http://www.uncsd2012.org/content/documents/GEA__concept_note_3March_references_01.pdf. Accessed 15 Aug 2013

FAO (2012b) Greening the economy with agriculture. Extract from the FAO council document CL 143/18: Status of preparation of FAO contributions to the 2012 united nations conference on sustainable development: governance for greening the economy with agriculture. http://www.fao.org/docrep/015/i2745e/i2745e00.pdf. Accessed 22 March 2017

FAO (2013) Climate smart agriculture sourcebook. http://www.fao.org/publications/card/en/c/6f103daf-4cd2-5a95-a03c-3d5d6b489fff. Accessed 21 March 2017

FAO (2014). Building a common vision for sustainable food and agriculture: principles and approaches. http://www.fao.org/3/a-i3940e.pdf. Accessed 27 Sept 2018

FAO (2015) FAOSTAT. http://faostat3.fao.org/home/E. Accessed 14 Nov 2015

FAO (2016) Increasing the resilience of agricultural livelihoods. http://www.fao.org/3/a-i5615e.pdf. Accessed 27 Aug 2018

FAO (2017) Land use, irrigation and agricultural practices – definitions. www.fao.org/fileadmin/ …/ess/…/Definitions/Land_Use_Definitions_FAOSTAT.xlsx. Accessed 15 Aug 2018

FAO (2018) Transforming food and agriculture to achieve the SDGs: 20 interconnected actions to guide decision-makers. http://www.fao.org/3/I9900EN/i9900en.pdf. Accessed 15 Aug 2018

Filmer D, Fox L (2014) Youth employment in Sub-Saharan Africa. Africa development series. World Bank, Washington, DC. https://doi.org/10.1596/978-1-4648-0107-5

IAASTD (2015) Agriculture at a crossroads. https://unwgfoodandhunger.files.wordpress.com/ 2015/09/agriculture-at-a-crossroads_global-report-english.pdf. Accessed 21 April 2017

Ikerd J (1993) The need for a systems approach to sustainable agriculture. Agr Ecosyst Environ 46:147–160

IPCC (2007) In: Metz B, Davidson OR, Bosch PR, Dave R, Meyer LA (eds) Climate change 2007: mitigation. Contribution of working group iii to the fourth assessment report of the intergovernmental panel on climate change. Cambridge University Press, Cambridge and New York

Jensen ES, Peoples MB, Boddey RM, Gresshoff PM, Henrik H-N, Alves BJR, Morrison MJ (2012) Legumes for mitigation of climate change and the provision of feedstock for biofuels and biorefineries a review. Agron Sustain Dev 32:329–364

Kirchmann H, Thorvaldsson G (2000) Challenging targets for future agriculture. Eur J Agron 12:145–161

Lipper L, Zilberman DA (2018) Short history of the evolution of the climate smart agriculture approach and its links to climate change and sustainable agriculture debates. In: Lipper L, et al. (eds) Climate smart agriculture: building resilience to climate change. Natural resource management and policy, vol 52. https://doi.org/10.1007/978-3-319-61194-5_2

Lopes C (2015) Agriculture as part of Africa's structural transformation. J Afr Transform 1(1), 2015, pp. 43–61 © CODESRIA & ECA 2015 (ISSN 2411–5002)

MEA (2005) Ecosystems and human well-being: synthesis. World Resources Institute, Washington, DC

Moodley S (2013) By declaring 2014 the 'year of agriculture', the African Union hopes to spur a green revolution. http://www.engineeringnews.co.za/article/by-declaring-2014-the-year-of-agriculture-the-african-union-hopes-to-spur-green-revolution-2013-09-27. Accessed 08 Oct 2018

Ocampo JA, Cosbey A, Khor M (2011) The transition to a green economy: benefits, challenges and risks from a sustainable development perspective. Report by a panel of experts to second preparatory committee meeting for united nations conference on sustainable development. United Nations Division for Sustainable Development (UN-DESA), United Nations Environment Programme and United Nations conference on trade and development

OECD (2011) Towards green growth. https://www.oecd.org/greengrowth/48224539.pdf. Accessed 15 Aug 2018

Pearce D, Markandya A, Barbier EB (1989) Blueprint for a green economy. Earthscan, London

Pretty J, Smith G, Goulding KWT, Groves SJ, Henderson I, Hine RE, King V, van Oostrum J, Pendlington DJ, Vis JK, Wlater C (2008) Multi-year assessment of Unilever's progress towards agricultural sustainability indicators, methodology and pilot farm results. Int J Agric Sustain 6(1):37–62

Power AG (2010) Ecosystem services and agriculture: tradeoffs and synergies. Philos Trans R Soc B 365:2959–2971

Schaffnit-Chatterjee C, Kahn B, Schneider S, Peter M (2011) Mitigating climate change through agriculture: an untapped potential. Deutsche Bank Research, Frankfurt

Smith P, Clark H, Dong H, Elsiddig EA, Haberl H, Harper R, House J, Jafari M, et al (2014) Chapter 11 - agriculture, forestry and other land use (AFOLU). In: Climate change 2014: mitigation of climate change. IPCC working group III contribution to AR5. Cambridge University Press, Cambridge

UN (2010) General Assembly, Report submitted by the Special Rapporteur on the right to Food, Olivier De Schutter Report A/HRC/16/49. https://www2.ohchr.org/english/issues/food/docs/a-hrc-16-49.pdf. Accessed 05 Dec 2018

UNDESA (2009) A global green new deal for sustainable development. United Nations Department of Economic and Social Affairs, New York

UNEP (2011) Towards a green economy: pathways to sustainable development and poverty eradication. United Nations Environment Program. ISBN: 978-92-807-3143-9 www.unep.org/greeneconomy. Accessed 02 Sept 2018

Wik M, Pingali P, Broca S (2008) Background paper for the world development report 2008: global agricultural performance: past trends and future prospects. World Bank, Washington, DC

Woldemichael AD, Salami A, Mukasa A, Simpasa A, Shimeles A (2017) Transforming Africa's agriculture through agro-Industrialization. Afr Econ Brief 8:7 (African Development Bank Group)

World Bank (2007) World development report 2008: agriculture for development. The World Bank, Washington, DC

World Bank (2012) Inclusive green growth: the pathway to sustainable development. The World Bank, Washington, DC

World Bank (2013) Growing Africa: unlocking the potential of agribusiness. AFTFP/AFTAI Report, The World Bank, Washington, DC

World Bank (2016) Poverty and shared prosperity 2016: taking on inequality. The World Bank, Washington, DC. https://doi.org/10.1596/978-1-4648-0958-3

World Bank (2018) https://data.worldbank.org/indicator/AG.LND.AGRI.ZS?view=chart. Accessed 15 Aug 2018

WWF (2017) Time is ripe for change: towards a common agricultural policy that works for people and nature. WWF Position paper. http://d2ouvy59p0dg6k.cloudfront.net/downloads/wwf_position_paper_on_cap_post_2020___final__contact_.pdf. Accessed 28 Sept 2018

Chapter 2
The Socio-Economic Context of Green Economy Implementation in the Agriculture Sector

2.1 Socio-Economic Considerations Related to Green Economy Implementation in a Developing Country Context

As discussed in Chap. 1, agriculture is a sector that directly affects human well-being in many ways. Green economy proponents suggest that there are numerous socio-economic gains that can be realised from adopting a green economic approach. Indeed, a number of the green economy principles highlighted in Chap. 1 are socio-economic in nature. For example, job creation, particularly 'green' job creation, is one of the principles of a green economy. Oxfam (2010) links the concept of green jobs to creating opportunities for vulnerable communities; as well as building resilience into the job market (for example the development of new employment opportunities and related skills requirements), resilience against climate change, as well as resilience within communities.

For a green economy, protection of livelihoods is important, not only through building internal resilience of agricultural communities to factors such as climate change, but also to global factors such as distortions in international trade. Equity (including intra-generational equity, both within and between countries; as well as inter-generational equity) is also a factor in green economy implementation. In a developing country context, an important consideration is how green economy projects in the agriculture sector can be implemented in order for the potential socio-economic benefits to be realised; and in particular, for those in greatest need to reap the benefits. According to UNEP (2011), a green economy in developing countries can take advantage of new growth trajectories that are socially inclusive, responsive to poverty eradication, and that respond to economic diversification; while also adhering to other green economy principles. These trajectories point to three essential actions: the enhancement of livelihoods, the creation of employment opportunities, and the reduction of poverty. Green economy project implementation has to be cognisant of these essentials.

© The Author(s), under exclusive license to Springer Nature Switzerland AG 2018
C. Musvoto et al., *Green Economy Implementation in the Agriculture Sector*,
SpringerBriefs in Agriculture, https://doi.org/10.1007/978-3-030-01809-2_2

Agriculture performs a range of interlinked functions for humans, spanning the social, economic and environmental domains; including food production, provision of environmental services, livelihood and economic opportunities. Livelihoods, employment and poverty are interlinked and feed into one another in complex ways, which makes it difficult to attend to only one of them at a time. The World Bank notes that while there has been much progress worldwide to curb the spread of extreme poverty, it remains unacceptably high for developing countries, especially those located in Sub-Saharan Africa (World Bank 2016a). Sub-Saharan Africa is home to 389 million people who are classified as "extremely poor" (World Bank 2016a). Green economy projects in Africa have to contribute to addressing this poverty.

The International Labour Organisation (2018) reports that globally, the total number of unemployed people in 2018 will rise above 190 million; while in developing countries the number is expected to increase by half a million per year in 2018 and 2019 respectively. Green economy projects should therefore endeavour to create employment. Keeping in mind that the green economy principles speak not only to green jobs, but also to decent work, the issue is not only about employment in general, but also about the kinds of employment opportunities created. Since 2012, there has been a steady rise in the number of workers in vulnerable positions across the globe. The ILO (2018) reports that 42% of the workers (1.1billion) worldwide are in a vulnerable form of employment, rising to 76% for workers in developing countries.

Unemployment and poverty are also often directly related to inequality; as addressing inequality generally also addresses poverty and employment. According to the World Bank (2016a), the general trend is that inequality is on the decrease. However, in many developing countries, the inequality gap is widening. It is within this context that green economy implementation in the developing world will take place. How green economy implementation addresses socio-economic considerations in a developing country context can be informed by the experiences of countries that have made progress toward a green economy. Botswana, for example, has identified socio-economic issues that pertain to a green economy in a developing country context; and these include strengthening social protection systems; making necessary changes to policies, institutions, regulations and incentives to address failures and facilitate participation; and capacity building, skills transfer and training (Green Economy Coalition 2012). The main requirement is to create an enabling environment that helps a developing country define its own path towards a green economy based on national circumstances, context and priorities.

2.2 Green Economy Implementation in Africa Within a Global Context: Commodification of Natural Resources and Financialisation of Crop Prices

The green economy is a broad and fluid concept, and this opens it up to different interpretations, definitions and practices (Bergius et al. 2018). In a world where the commodification of nature is increasing, for example emissions trading and the trade in drinking water, there have been perceptions that the green economy is an attempt to 'commercialise' and 'commodify' nature (e.g. Unmüßig et al. 2012; Levidow 2014). The fluid nature of the green economy concept is evident in the way its implementation has been interpreted in the agriculture sector. According to Bergius et al. (2018), the green economy label has been applied to a wide range of different practices, from the greening of current neoliberal economies,[1] to radical transformation of these economies. Buseth (2017) and Bergius et al. (2018) report that agricultural investment programmes have been presented as green economy implementation, an example being the Southern Agricultural Growth Corridor of Tanzania (SAGCOT). SAGCOT is championed at the global level as a good example of how the green economy can be implemented in practice (Buseth 2017). However, Buseth (2017) argues that SAGCOT's policy is not a good representation of inclusive green growth, but rather of the global rush for land and of the agri-business sector finding new, attractive labels within which to frame their interest and investments. The trend of land acquisition under the green economy banner has also been highlighted by Nhamo and Chekwoti (2014).

In addition to commodification of natural resources, another phenomenon that has implications for green economy implementation is the financialisation of agricultural commodities. In the context of agriculture, Ouma (2014: 163) defines financialisation as the "more general penetration of food production and agro-food chains by 'finance capital'". Krippner (2011: 4), describes the phenomenon as "the tendency for profit making in the economy to occur increasingly through financial channels rather than through productive activities." According to Staritz et al. (2015), financialisation is connected to increased activity in international commodity exchanges driven purely by financial interests.

Globalisation and financialisation play a role in determining global commodity prices. Crop prices are determined through world markets and commodity index funds (Peralta 2017; Ederera et al. 2016). While various factors including supply and demand play an important role in determining commodity prices, there has been debate about the additional role of financial investors, and how their increasing presence in commodity derivative markets affects commodity prices (Ederera et al. 2016). Peralta (2017) notes that in agriculture, financiers and speculators are turning to food crops and mono-cultural farming estates as lucrative areas for profit-making; which has resulted in unstable food prices and large-scale land acquisitions.

---

[1]Neoliberalism is a term commonly used to describe free-market economics. It involves policies associated with free trade, privatisation, price deregulation, a reduced size of government and flexible labour markets (Pettinger 2018).

Crop prices have a significant impact on farmers (both as producers and consumers of agricultural commodities), and this impact can be severe for small-scale farmers in Africa, who have little if any way of protecting themselves from what happens in world markets. Large international and financially adept actors stand to gain from opportunities for speculation and hedging activities on derivative markets, while local actors in producer countries face greater challenges in an environment of price instability and short-termism (Newman 2009). Peralta (2017) notes that globalisation has drastically changed the agricultural landscape in recent decades, with severe social and ecological impacts, including hunger, displacement, and the pollution of soil and water.

International agricultural initiatives (whether labelled as green economy initiatives or not) that aim to benefit small-scale farmers do not always achieve this goal. Documented experiences from Mozambique, Tanzania and the Philippines highlight that agricultural programmes supported by some international organisations have reportedly contributed to a number of negative consequences (Peralta 2017). Examples of these consequences include the relocation of farmers to semi-arid or less fertile lands, precarious employment for farmers as plantation workers and contractors, indebtedness and increasing poverty (Peralta 2017; Bergius et al. 2018).

Opportunities for green economy implementation through international agricultural initiatives should not only be perceived in a negative light. Such initiatives, if properly managed, could provide prospects for transforming farms from generally precarious subsistence operations to more commercial and sustainable agrienterprises. What is required are mechanisms to ensure that developing countries and smallholder farmers benefit most; and that the vulnerabilities of farmers are not exploited. Green economy implementation therefore has to put in place safeguards for vulnerable farmers. Displacement of farmers and creation of large-scale enterprises is not a necessity for green economy implementation. Promotion of the green economy in agriculture should be about enhancing agricultural practices and the general agricultural environment to make them not only compatible with the values of the green economy, but to make them more socially, economically and environmentally beneficial for farmers and developing countries. A green economy that is beneficial requires enabling policies and strategies that correctly interpret and apply the green economy and its ideals. If the green economy is to address the challenges faced by farmers effectively, especially in Africa, it should be implemented in ways that address the complex socio-economic issues that affect agriculture and not be used as a conduit for further globalising capitalism. Investments into green economy implementation in the agriculture sector should be made in ways that directly benefit farmers and enhance their control over agricultural production, rather than the opposite.

2.3   World Trade in Agricultural Commodities and the Small-Scale Farmer

Green economy implementation in any country would be affected by both local and global trade and market factors. The rapidly changing and globalising political economy of agriculture should be considered in green economy implementation. This is in light of the increasing role of trade in agriculture, market speculation, changing nutrition in emerging markets, food insecurity, land-grabbing, and climate change (Karapinar 2010). Trade in agricultural commodities is crucial for the livelihoods of many small-scale farmers. Tens of millions of small-scale producers and farm workers earn their living from growing crops for export (Burnett and Murphy 2014). Pirnea et al. (2013) note that the economic viability of many small farms depends on income from sales of internationally traded commodities. Furthermore, world trade in agricultural commodities also affects the prices of commodities sold on local markets.

Otte (2007) asserts that globalisation and the associated liberalisation of world trade are widely perceived as a major threat to developing countries in general and to smallholder farmers in particular. There are concerns about the implications of globalisation for small, vulnerable subsistence producers (von Braun and Meinzen-Dick 2009); particularly in the face of rapid changes in global trade in agricultural commodities in terms of size, the way it is organised, the issues that it is concerned with and the main countries involved (Swinnen et al. 2013; Burnett and Murphy 2014; Piñeiro and Piñeiro 2017). Increasing trade is taking place between developing countries, and environmental concerns have assumed increasing importance in agricultural trade matters (Piñeiro and Piñeiro 2017). All these factors have implications for green economy implementation in the agriculture sector. Green economy projects thus have to be conceived and implemented in a way that ensures that they can navigate these global issues and operate sustainably.

International trade in agricultural commodities is increasingly taking place through high value Global Value Chains (GVCs) (Swinnen et al. 2013; Burnett and Murphy 2014). GVCs are characterised by the dominance of a few large multi- and transnational food corporations and the application of rigorous safety and quality standards with respect to marketing, labelling, food contamination, hygiene and traceability (Swinnen et al. 2013). According to Swinnen et al. (2013), such standards increasingly include ethical and environmental standards as well. Trade through GVCs has implications for farmers around the world, who find themselves confronted with new competitive pressures, as well as new opportunities (Swinnen et al. 2013). For small-scale farmers, participation in GVCs can have advantages such as access to capital, technology, markets, enhanced skills and management techniques (UNCTAD 2009). Small-scale farmers can participate in GVCs through contract schemes where farmers are provided with business development services such as inputs, technical assistance, and credit by private sector actors (both domestic and/or international). In return, farmers commit to sell their output to these providers, subtracting the cost of the supplied inputs from their total profits (von Braun and Meinzen-Dick 2009).

However, integration into GVCs also has potential risks for farmers. The integrated nature of GVC trade can amplify shocks, contributing to instability of output and employment in GVCs (TRALAC 2016).

Concerns have been raised that transnational corporations abuse their market power, add downward pressure on rural wages and disempower farmers through unfair contractual arrangements (FAO 2003). There are also perceptions that in a global agricultural economy, large farms controlled by giant multinational corporations will continue to displace smaller farms in the global marketplace (Pirnea et al. 2013). However, there are reported efforts by multinational firms to increasingly include African smallholder farmers in their GVCs (Banerjee and Duflo 2011; Vermeire et al. 2017). While some success has been achieved with these efforts, the number of success stories is reported to be much less than the number of failures (Banerjee and Duflo 2011; Vermeire et al. 2017). Although African smallholder farmers participate in GVCs to varying degrees, Lutz and Olthaar (2017) point out that participation in GVCs on its own is not sufficient; and it is critical to understand whether smallholders are able to create a competitive advantage and appropriate a reasonable share of the value created.

Trade distortions are a reality in the agriculture sector; and green economy implementation in the sector takes place in a context of difficulties in setting rules and deadlocks over trade. The ILO and UNCTAD (2013) note that agriculture is among the most distorted sectors in international trade, with relatively high tariffs and subsidies, which are not allowed in other sectors. According to Elliott (2018), in the early 2000s, subsidies and trade barriers in rich countries were helping drive down agricultural prices, leaving poor farmers in developing countries struggling to support their families. Elliott (2018) notes that although the importance of agriculture in overall international trade has declined, it is still of critical importance for developing countries, especially in Africa, for both food security and livelihoods, and particularly for vulnerable groups such as women.

As green economy implementation takes place in a rapidly changing global context for the agriculture sector; it should be cognisant of issues on global agricultural commodity markets and how they affect African farmers. Failure to proactively deal with the consequences of changing dynamics in markets could have devastating effects on the well-being of farmers implementing agricultural green economy projects. Green economy initiatives have to appropriately respond to global dynamics in the production and marketing of agricultural commodities; should not exacerbate the problems faced by African farmers; and should endeavour to address the problems and to optimise their participation in agricultural trade at both the local and global levels.

2.4   Green Economy Implementation in a Rapidly Changing Technological Context: Disruptive Technologies in Agriculture

Technology is changing every aspect of human life, and agriculture is no exception. Disruptive technologies, generally defined as new ways of doing things that disrupt or overturn existing practices and/or technologies (Christensen 1997), are a reality in every sphere of human endeavour. According to the World Economic Forum (WEF 2018), emerging technologies driven by the Fourth Industrial Revolution (4IR) are disrupting many industries, bringing rapid and large-scale change, Disruptive technologies have also been driving innovations in the agriculture sector (Hall and Martin 2005).

Emerging technologies include digital building blocks such as big data, the Internet of Things (IoT), artificial intelligence, autonomous vehicles, advanced robotics (WEF 2018), and drones. King (2017) notes that a technological revolution in farming led by advances in robotics and sensing technologies is set to disrupt modern practice. 4IR technologies have the potential to help revolutionise food systems, dramatically changing the shape of demand, improving value-chain linkages and creating more effective production systems (WEF 2018). IoT can be applied to agriculture in order to reduce production costs and increase product quality. The use of drones in agriculture is on the increase; with significant potential for supporting evidence-based planning and spatial data collection (FAO and ITU 2018). Drones also have major application for precision farming, such as in fertiliser application, irrigation and precision spraying (FAO and ITU 2018); thereby contributing to more precise and effective farm management systems (AU and NEPAD 2018).

Emerging technologies can contribute towards meeting the principles of a green economy. Devices such as robots and drones, could allow farmers to slash agrichemical use by spotting crop enemies earlier, allowing precise chemical application or pest removal (King 2017). Such approaches would improve targeting of agrichemicals, reduce the negative environmental impacts of agrichemicals and cut costs. In a green economy context, this would be aligned with green economy principles focusing on environmental protection and resource efficiency. These technologies can be used in small scale farming and therefore in associated green economy initiatives. Use of drones in small-scale farming in Mozambique is reported to have helped farmers make decisions that improved crop water use efficiency and yields (AU and NEPAD 2018).

Guilleń-Navarro et al. (2017) argue that agriculture could apply IoT to transform towards increased automation, reduced production costs, increased production and quality, and improved responsiveness to unfavourable weather and climatic conditions. This would have both advantages and disadvantages in a green economy context. Increased automation could reduce low skills employment opportunities and this would compromise the attainment of green economy ideals focused on improving human well-being. Reducing costs and increasing production and quality would be aligned with resource efficiency green economy aspirations.

The WEF (2017) notes that food systems will be dramatically influenced by new technologies; and that such technologies depend on a world increasingly connected to the internet. The internet penetration rate is relatively low in Africa, with an average rate of 35.2% in December 2017; and varying between countries, ranging from as low as 4.3% in Niger, to 68% in Tunisia (Internet World Stats 2018). Furthermore, gender inequality and geography compound the internet access challenge. The World Bank (2016b) points out that those without access to internet are predominantly rural, poorly educated, with lower incomes, and a large number are women and girls. All these factors would hamper the application of emerging technologies to green economy projects in the agriculture sector in rural areas in most African countries.

While 4IR technologies have the potential to help revolutionise food systems in positive ways, these technologies are likely to introduce new challenges pertaining to health and safety, the environment, privacy and ethics (WEF 2018). The technologies will raise a new set of social questions, relating to the control of data, the future of jobs, and the role of technology in food production (WEF 2017). WEF scenarios (WEF 2017) demonstrate that technology has the potential to exacerbate inequality if not directed with purpose at the needs of a global population. These concerns are all central to the green economy, and implementation has to take this into consideration and ensure that technologies align with green economy ideals, while ensuring that agriculture remains competitive, sustainable and profitable. Specific issues related to new technologies also have to be addressed in green economy project implementation in specific contexts. The deployment of drone technology in Africa, for example has technological, economic, social, legal and regulatory challenges (AU and NEPAD 2018). Furthermore, the use of new technologies requires specific skills, ranging from piloting drones to operating geographic information systems (GIS) and data analysis software, interpreting data, and providing agronomic or spatial planning advice (AU and NEPAD 2018).

The rapidly changing technological context has implications for green economy implementation. Implementers have to keep abreast of new and emerging technologies; and be prepared to adopt them and to adapt agricultural practices to work with the technologies. This is important for maintaining the competitiveness of agriculture, as commodities produced in Africa and other parts of the developing world have to compete with commodities produced more efficiently elsewhere. The commodities would also have to meet high consumer quality requirements, which can be difficult without the aid of technology. Green economy implementation should be cognisant of the fast changing technology landscape and be ready to adopt and use new technologies. New technologies should not be adopted blindly but should be assessed in the context of a green economy and potential negative impacts such as loss of employment associated with some technologies should be addressed.

2.5 Implications of Small-Scale Farming and Poverty for Green Economy Implementation

Farming, particularly small-scale farming, has an important role to play in the world's efforts to curb poverty. According to the World Bank (2017), agricultural development is an effective mechanism to combat the spread of extreme poverty, enhance shared prosperity, and decrease hunger, by feeding an estimated 9.7 billion people by 2050. One of the ways in which agriculture does this is by being a major employer, especially for people in developing countries, and as such is an essential source of income and livelihoods for the poorest of the poor. In fact, the poverty profile developed by the World Bank in 2016 suggests that the global poor are mainly rural, young, poorly educated, predominantly employed in the agricultural sector, and live in large households with relatively high numbers of children (World Bank 2016a) (Box 2.1). In the context of South Africa and Africa, where rural poverty is high, agriculture is particularly relevant for a green economy which addresses rural poverty. Green economy implementation in such a context has to accommodate the needs of the poor and has to include mechanisms to ensure their sustained participation in projects .

Green economy implementation in the agriculture sector has to occur within a specific context and cater for differences between and within countries. The situation in South Africa, for example is different from that in most African countries, where small-scale farmers produce most of the food. In South Africa large-scale commercial farmers produce around 95% of the agricultural output, with the smallholder sector producing 5% (Aliber and Hart 2009).

In 2017, close to 70% of the population in Africa was involved in agriculture as smallholder farmers working on small parcels of land (average of less than 2 hectares) (AGRA 2017). According to ASFG (2010), agriculture is the lifeblood of African economies and societies, with some 65% of the population (more than 80% in some countries), depending on small or micro-scale farming as their primary source of livelihood. However, the degree of dependence on small-scale agriculture varies substantially across and within countries in Africa (Gollin 2014). In some East African countries (Kenya, Tanzania, Ethiopia and Uganda), smallholder farming accounts for about 75% of agricultural production and over 75% of employment (Adeleke et al. 2010). In South Africa on the other hand small-scale agriculture is not the main source of livelihoods for rural people (Kingdon and Knight 2004). Despite the dominance of the large-scale commercial sector in terms of land area, volume

Box 2.1 Poverty profile of the world in statistics

Poverty profile of the world in statistics:
80% of world poor live in rural areas,
64% work in agriculture,
44% are 14 years or younger, and,
39% have no formal education.

Source: World Bank 2016a

and value of outputs, the South African government recognises the potential role of small-scale farming for the economy and job creation as has explicitly stated the need to encourage small-scale farmers to produce and drive economies in their respective communities (Gabara 2012). The importance of small-scale farming in food production and/or livelihoods in Africa means that green economy implementation in the agriculture sector has to focus on small-scale farming.

African countries, including South Africa, have committed to the principle of green economic growth, based on a green economy (UNEP 2013). They have also affirmed the key role of agriculture in their economies (Moodley 2013). Furthermore there has been recognition among African states of the need to ensure that the adoption of a green economy takes into account the particular social and development imperatives of African states, especially poverty reduction (African Union 2011). Agriculture is therefore central to Africa's green economic development and poverty eradication plans. Green economy implementation in the agriculture sector has to be cognisant of this situation and take measures to ensure that projects are designed in such a way that agriculture fulfils its envisaged role as a driver of green economic growth and poverty eradication. Small scale farming has to be at the centre of green economic development in most African countries and green economy implementation thus has to care of the specific needs of small scale farmers.

2.6 National and Local Development Considerations in Green Economy Implementation: The Case of South Africa

The green economy has to be relevant to the local context, and its implementation has to be informed by local realities. This section uses the South Africa context to illustrate how national and local development considerations impact the way in which green economy implementation would play out. According to the African Centre for a Green Economy (AFRICEGE 2015), South Africa faces development challenges which are directly associated with natural resource constraints such as arable land and water. This is linked to the high levels of unemployment, poverty and inequality in the country (AFRICEGE 2015). The adoption of a green economic approach is therefore seen as not only an avenue to achieve sustainable development, but also to address the developmental challenges of the country (AFRICEGE 2015).

South Africa's National Development Plan (NDP) (NPC 2011) articulates the country's general development and green economy aspirations, and should be a key consideration in green economy implementation. One of the aspirations in the NDP is that "by 2030, South Africa's transition to an environmentally sustainable, climate change resilient, low-carbon economy and just society will be well under way" (NPC 2011). In order for South Africa to achieve this vision, challenges such as poverty, inequality and unemployment need to be addressed. Currently, South Africa's unemployment rate stands at 26.7% (StatsSA 2018a), up from 24.9% in

2012. StatsSA (2017b) reports that poverty is on the rise in South Africa; with 55.5% of South Africans being classified as poor in 2015, a rise of about 2% since 2011. The NDP has a number of guiding principles for a smooth transition to a green economy which addresses developmental challenges (NPC 2011), and some of these include just ethical and sustainable approaches; ecosystems protection and acknowledgement that human well-being is dependent on the health of the planet. These principles should inform green economy implementation in South Africa.

The groups most vulnerable to poverty in South Africa are children (aged 17 years and younger), black South Africans, women, people from rural areas, citizens with no or little education, and people residing in the Eastern Cape and Limpopo provinces (StatsSA 2017a). In 2017, the unemployment rate for black South Africans stood at 30% and was higher than that for other population groups (StatsSA 2018b),[2] while in 2013 the national unemployment rate for females was 5.4 percentage points higher than the rate for males (StatsSA 2013). Young females are also particularly vulnerable to unemployment, with 35% of young women (between the ages of 15 and 24) not being in employment, education or training (NEET) whereas 29.6% of young men are in a similar situation (StatsSA 2018a). Reducing unemployment in general, and for groups experiencing the highest unemployment has to be a key consideration for green economy implementation in South Africa.

South Africa's New Growth Path (EDD 2011a), projects that the green economy will create well over 400 000 jobs by 2030, many of which will be classified as green jobs. According to the ILO (2016), green jobs are "jobs that contribute to preserving or restoring the environment, be they in traditional sectors or in new, emerging green sectors such as renewable energy and energy efficiency". Job creation in a green economy is therefore not about the creation of any job, but requires that the jobs created have specific attributes (such as those for green jobs) and are compatible with green economy principles.

Women are more likely than men to be poor in South Africa. According to StatsSA (2017b), 57.2% of women live below the upper-bound poverty line (UBPL), in comparison with 51.4% of men. Addressing female poverty in South Africa through the green economy in an agricultural context requires gender mainstreaming. Babugura (2017) argues that it is vital to appreciate the critical need for gender equality in a green economy due to the fact that the economy and labour market do not provide equal opportunities for women and men. Gender-responsive policies and programmes should therefore be embedded in and be integral to a green economic approach. Implementing the green economy without such policies and programmes might exacerbate gender inequalities, thus thwarting the overall rationale of sustainable development (Babugura 2017).

The youth of South Africa is one of the groups that need support through national development and growth plans. In 2018, 3.3 million out of 10.3 million young people between the ages of 15 and 24 were not in employment, education or training (StatsSA

---

[2]This definition of unemployment considers a person to be unemployed only if they have made some sort of effort to find employment or to start something themselves in the four weeks prior to them reporting their status, in other words they are searching for employment.

2018a). Youth employment and skills development are flagged in national strategies such as the Green Economy Accord (EDD 2011b) and should be addressed in green economy implementation. Mudombi (2017) argues that the youth, while relatively inexperienced, are ideally situated to benefit from jobs created in the green economy. While many of the envisaged green jobs might require new and different skills, the youth have the capacity to easily adapt to change, are willing to experiment and more likely to learn a new skill, thus making them more adaptable to the new 'green' job requirements. The green economy can play an important role in the reduction of youth unemployment and in the upskilling of the youth in South Africa. This will require a multi-modal approach which includes support and incentives such as targeted educational programmes, enabling youth access to finance and an appropriate and enabling legislative framework (Aceleanu et al. 2015).

In South Africa's current context, it is important to ensure that green economy implementation does not exacerbate the skills, employment and opportunities constraints as well as the related inequalities along racial, gender and age group lines. The South African example illustrates how green economy implementation is not going to be a 'one size fit all', despite its universal principles. As South Africa's particular developmental context places specific requirements on green economy implementation, other countries are likely to have different contexts and developmental issues and these would affect green economy implementation differently.

2.7   Governance and Policy Issues in Green Economy Implementation—the Case of South Africa

Countries that promote a green economy driven by agriculture have to create an environment that enables both agriculture and the green economy to thrive. An appropriate enabling environment is crucial for green economy implementation and for achievement of its ideals. Suitable policies, legislation and governance arrangements are at the core of an enabling environment; and these should be underpinned by frameworks of regulation, incentives and disincentives that account for the all-encompassing nature of a green economy in an agricultural context. An appropriate environment has to ensure that practices across the board are aligned to green economy principles and meet agriculture's objectives. A green economy is a multi-stakeholder endeavour which is only achievable through the actions of individuals and groups from multiple sectors, including government, business and civil society (Stafford et al. 2014). This has to be reflected in legislation, policies and governance arrangements.

Land governance and tenure security are key factors in creating an enabling environment for a green economy. Tenure security can be defined as people's ability to control and manage land, use it, dispose of its produce, and engage in transactions, including transfers (IFAD 2015). Tenure security influences the extent to which farmers are prepared to invest in improvements in production and land management, and helps ensure access to the natural resources that are needed for farming (Sustainable

Development Solutions Network 2014; IFAD 2015). In the context of a green economy, UN Environment argues for policies that support improved land tenure rights of smallholder farmers as more secure tenure is needed if farmers are to take on more risks associated with embarking on new green agricultural initiatives (UNEP 2011). In addition, improvement of land tenure rights has to address the needs of women, due to their central role in agriculture in Africa. Women produce an estimated 60–80% of food in developing countries, yet they rarely have secure land rights (Sustainable Development Solutions Network 2014).

In South Africa, security of tenure is precarious for many rural residents (Africa Research Institute 2013; Clark and Luwaya 2017). Legislation designed to improve rights of tenure, for example the Extension of Security of Tenure Act (Act No 62 of 1997) (Republic of South Africa 1997), has been ineffective and in some instances poorly enforced (Africa Research Institute 2013). In response, the government of South Africa is in the process of instituting land reform. One of the components of the land reform policy is tenure reform, which is aimed at protecting, securing and strengthening the rights that people have over land (Clark and Luwaya 2017). Implementing green economy projects requires some long-term investments. Such investments could be risky, particularly for smallholder farmers, who can often only afford to adopt short-term survival strategies. Long-term security of tenure must be in place for farmers to have an incentive to undertake such investments.

In a green economy, actions taken to reach economic ends should advance social and environmental ones, just as actions taken to meet social and environmental ends should strengthen and develop the economy (Halle 2011). This would have to be reflected in green economy legislation, governance and policies at different levels (national, local and sector). To guard against reverting to entrenched practices of pursuing economic objectives at the expense of the environment and society, the green economy governance environment should include incentives and disincentives for ensuring that implementation is aligned with green economy ideals. According to Halle (2011), incentive and disincentive measures should be empowered by the right mix of legislation, institutional tracking mechanisms, third party monitoring, and funding mechanisms to allow rewards to be offered and legal mechanisms for sanctioning if required.

South Africa has adopted an inclusive approach to the green economy, with various stakeholders involved. An example of this approach is the country's Green Economy Accord (EDD 2011b), which is an agreement involving business, labour, community and the state to shift the economy to a greener and more labour-absorbing trajectory. The inclusive approach is also reflected at government level. The implementation of South Africa's green economy is decentralised and includes all levels of government (DEA 2017). Overseeing green economic development in South Africa is a co-responsibility of the Economic Development Department (EDD) and the Department of Trade and Industry (DTI), with a number of other government departments also playing a role (Fig. 2.1).

South Africa has not enacted specific legislation for the green economy, and there is no integrated policy response to support green economy implementation. However, the country has a number of policies and strategies that are aligned with the

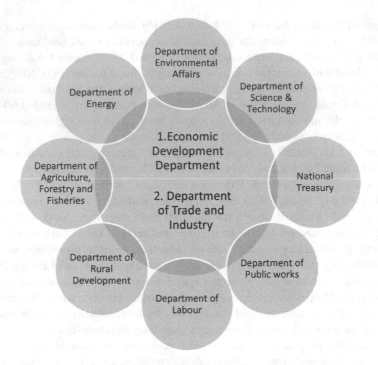

Fig. 2.1 South Africa's green economy governance arrangements are based on partnerships at national government level (Derived from DEA 2017)

green economy. According to the Partnership for Action on Green Economy (PAGE 2017), South Africa's vision of transitioning to a greener economy is supported by an extensive policy and regulatory framework. Many of the policies and strategies were put in place to support the country's sustainable development aspirations, and originated before the green economy concept came into widespread use and its formal adoption by the government of South Africa. These policies and strategies are also relevant to agriculture and the green economy. Within the existing framework, the key national strategies and plans that are explicitly relevant to the green economy in general, including issues of significance to agriculture, include the National Framework for Sustainable Development (NFSD) (DEAT 2008); the National Strategy for Sustainable Development and Action plan (NSSD) (DEA 2011); the New Growth Path, specifically the Green Economy Accord (EDD 2011b); the Medium Term Strategic Framework (Presidency Republic of South Africa 2009) and the National Development Plan (NPC 2011).

In addition to national strategies and plans, various green plans exist at the provincial and local levels; for example for the provinces of Gauteng (Gauteng Province 2011), Western Cape (Western Cape Government 2013), KwaZulu-Natal (DEDT KwaZulu Natal 2012); and the City of Tshwane (City of Tshwane 2013). Table 2.1

provides a summary of the provincial and local level strategies, plans and programmes of relevance to the green economy in South Africa.

Although there are various government policies of relevance to green economy implementation, there is no system for ensuring consistency. Swilling et al. (2016) note that there is no planned centre within government that creates coherence among the various components of South Africa's green economy policy framework. The existence of numerous policies and strategies could be a source of confusion through policy incoherence. According to Swilling et al. (2016) and Nicholls et al. (2016), policy incoherence could be an obstacle to transitioning to a green economy in South Africa. Swilling et al. (2016) highlight that there are important policy gaps and a lack of integration of policies across different government departments and spheres in South Africa.

While the central role of agriculture in South Africa's green economy is acknowledged in strategies such as the Green Economy Accord (EDD 2011b), there is no cohesive framework for agriculture in the green economy. However, there are agricultural laws and policies that align with the green economy, in particular legislation pertaining to sustainable agriculture and protection of natural resources, and safeguarding of human well-being. Relevant legislation and policies include:

- The Conservation of Agricultural Resources Act (CARA) 43 of 1983 (Republic of South Africa 1983) which deals with the conservation of natural resources for the purposes of agriculture.
- Strategic Plan for South African Agriculture (NDA 2002), which aims to increase wealth creation in agriculture and rural areas, sustainable employment in agriculture, farming efficiency, national and household food security, and incomes and foreign exchange earnings from agriculture.
- The draft Sustainable Utilisation of Agricultural Resources Bill of 2003 (NDA 2003) provides for the optimum productivity and sustainable utilisation of natural agricultural resources and biodiversity protection.
- The 1998 Agricultural Policy for South Africa (NDA 1998) addresses the protection of the natural resource base, prevention of degradation of soil and water, and conservation of biodiversity. It also provides for contribution to the economic and social wellbeing of all, ensuring a safe and high-quality supply of agricultural products, and safeguarding the livelihood and wellbeing of agricultural workers and their families (Scotcher 2009).

Although policies and legislation specifically designed for green economy implementation in a sector such as agriculture can facilitate implementation and provide certainty, a lack of such policies does not necessarily derail implementation. Existing aligned policies and governance arrangements could be used to provide the legislative and policy guidance required before such time as sector-specific green economy policies and legislation are developed.

In addition to government strategies, there are organisations in South Africa (including the private sector, civil society and farmers' groups) which are running and/or supporting agricultural initiatives with some alignment to green economy principles. For example, Woolworths Holdings Limited (WHL), a major national food

Table 2.1 South Africa's plans and programmes of relevance to the green economy in an agricultural context

Strategy/plan or programme	Responsible agency	Core issues addressed
National Strategy for Sustainable Development and Action Plan	Department of Environmental Affairs (National)	• Sustainable water and land resources management • Protection of agricultural land • Sustained food security • Local economic development
Green Economy Accord	Economic Development Department (National)	• Unemployment, poverty and inequality • Support for small-scale agriculture • Agro-processing chains and expanding trade • Biofuel production
National Development Plan	National Planning Commission	• Land reform and security of tenure • Expansion of agriculture and improving efficiency of irrigation • Increase food production and raise rural incomes and employment opportunities • Infrastructure for agriculture and farmer support • Improving efficiency of irrigation
Medium Term Strategic Framework	The Presidency (National)	• A competitive economy • Decent work opportunities • Growth in core productive sectors including agriculture
Strategic plan for the Department of Agriculture, Forestry and Fisheries	Department of Agriculture, Forestry and Fisheries (National)	• Supporting subsistence farmers and small-scale producers • Promoting agro-ecological agriculture • Efficient use of natural resources • Protection of indigenous genetic resources • Green jobs to improve livelihoods • Food security, agrarian reform and profitable food production
Green economy strategy for KwaZulu-Natal Province	KwaZulu-Natal Provincial government	• Green jobs • Self-sufficiency (produce own food, water and energy) • Comprehensive overhaul of the whole economy
Green strategic programme for Gauteng	Gauteng Provincial government	• Food security • Local organic production • Small scale urban agriculture

(continued)

Table 2.1 (continued)

Strategy/plan or programme	Responsible agency	Core issues addressed
Green is smart: Western Cape green economy strategy framework	Western Cape Provincial government	• Support for agri-production and expanding value chains and markets • Sustainable farming practices • Energy and water efficiency • Waste beneficiation • Food security
Green economy strategic framework for the City of Tshwane	City of Tshwane Metropolitan Municipality	• Promote sustainable agriculture and agro-ecology • Rehabilitate degraded lands and promote their sustainable use • Promote small-scale organic farming, community co-operatives and local food markets • Promote urban agriculture • Support programmes for protection of agricultural land, sustained food security and local economic development
Nelson Mandela Bay climate change and green economy action plan	Nelson Mandela Bay Municipality	• Reduce food insecurity • Develop commercial community food gardens • Create jobs through labour intensive agricultural activities • Productively use unutilized/underutilized land

retailer, runs the Farming for the Future auditing and certification scheme. Under the scheme, WHL works directly with 98% of their primary produce suppliers in South Africa to implement sustainable farming practices that use fewer natural resources, promote soil health and reduce use of synthetic fertilisers and chemicals (WHL 2018). Initiatives such as the WHL scheme provide a potential avenue for green economy implementation, and illustrate the diversity of stakeholders that could play a role in green economy implementation at project level, and also the range of scales at which such projects could be implemented. Policies are required to guide the participation of varied stakeholders in green economy implementation to ensure alignment and coordination among the different activities and actors.

2.8   Environmental Governance and Justice in the Context of a Green Economy and Agriculture

With the global movement towards sustainable development, and the increasing pressure on governments to find green alternatives to sustain their developmental trajectories, it has been suggested that the world's ability to sustain development will rest on two important yet converging threats to humanity (UNDP 2014). These threats are (i) the inescapable reality that while there may be a general trend of decreasing inequality (World Bank 2016a), for a large part of the world's population living in the developing world inequality is rising, and (ii) the increasing number and complexity of the risks that are arising from induced environmental change as we continue to surpass the Earth's planetary boundaries (UNDP 2014). These issues make environmental justice an integral part of development and green economy implementation.

According to UNDP (2014: 6), "environmental justice is about legal transformations aimed at curbing abuses of power that result in the poor and vulnerable suffering disproportionate impacts of pollution and lacking equal opportunity to access and benefit from natural resources". Examples of such abuses of power include profit and benefit accrual through the colonisation of ecosystem goods and services, as well as the unequal distribution of risk due to environmental hazards and climate variability. For example, the distribution of green space in cities has received attention the world over, as it has been a clear marker of inequality between rich and poor (Wolch et al. 2014; Haaland and van den Bosch 2015). Other examples include agriculture, in particular the juxtaposition of sustainable agriculture and the demands of the global food system (Gottlieb and Fisher 1996; Bradley and Harrera 2016); and fair trade practices through research on ethical production and consumption links through the reconnection of producers and consumers in a global moral economy (Goodman 2004; Marston 2013). The field of environmental justice includes the interplay between policies, regulations and economic context (Bryant 1995), as well as an approach that includes factors such as class, gender and citizenship; towards the development of society in a way that is sustainable, productive and equal regardless of race, class, gender or citizenship status (Berkey 2014; Pellow 2000; Sze and London 2008). For Berkey (2014: 10), environmental justice refers to "those cultural norms and values, rules, regulations, behaviours, policies, and decisions to support sustainable communities where people can interact with confidence that the environment is safe, nurturing, and productive.

The green economy, depending on how it is implemented, could contribute to environmental justice and equality. With the move to a green economy, much emphasis is placed on the 'transition' that is needed in most sectors, and in the economy as a whole. One way of ensuring a just transition is to move beyond how the green economy is defined in words, to taking actions that contribute to achieving social justice. Ehresman and Okereke (2015) argue that the green economy should be viewed through the lens of environmental justice, and that a better understanding of the justice implications of the green economy is required.

In order to fully benefit from a green economy it is necessary to assess and establish its relationship with social and environmental justice (Ehresman and Okereke 2015). Currently, there are a number of criticisms of the green economy; including that it may perpetuate previous systems that engender inequality and unequal access to resources, not only between individuals, but also between nations. As such, there is a need to "carefully consider if, where and how policies aimed at encouraging a greener economy can better take account of the full range of justice impacts and prospects such a transition would generate" (Ehresman and Okereke 2015: 9).

History has shown that small-scale farmers generally face a combination of environmental and economic problems that are fraught with entrenched marginalities and disparities. As such, it is imperative that small-scale farmers benefit from the guidance provided by environmental justice proponents. The green economy is an ideal mechanism to ensure environmental justice for small-scale farmers in Africa and the rest of the developing world, provided it is implemented in a way that ensures equality and rejects inequality in all its forms. By following this path, the green economy can contribute to ensuring the mobilisation and securing of rights, health and welfare, as well as quality of life for small-scale farmers.

2.9 Integrating Socio-Economic Green Economy Principles at Project Implementation Level

The basis of a green economy is enhancing human well-being without damaging the environment and this is highlighted in the way the green economy is commonly defined, e.g. UNEP (2012). It is clear from its principles (see Chap. 1) that a green economy aims to simultaneously address social, economic and environmental issues. A number of the green economy principles identified in Chap. 1 focus on issues of human well-being; and clarify the social and economic aspirations of a green economy, as well as issues related to governance and decision making. For example:

- The green economy should create decent work and green jobs;
- The green economy internalises externalities;
- The green economy is equitable, fair and just—between and within countries and between generations;
- The green economy delivers poverty reduction, well-being, livelihoods, social protection and access to essential services;
- The green economy improves governance and the rule of law. It is inclusive, democratic, participatory, accountable, transparent and stable;
- The green economy uses integrated decision making.
  Source: Allen (2012).

These principles highlight some of the socio-economic and governance considerations for green economy project implementation. These considerations have to be incorporated into projects in a way that makes it possible to account for and take actions that meet agricultural objectives. But what does this mean in practice in an agricultural context? For one thing, green economy projects have to adopt agricultural practices that not only ensure production, but enhance human well-being. Agriculture is the foundation of people's livelihoods and of economic development in many developing countries; providing food security, as well as other products such as energy and fibre. Agriculture also contributes significantly to employment and poverty reduction. This central role of agriculture in human well-being makes green economy projects in the sector well-suited to adopting socio-economic green economy principles.

To understand what it takes for an agricultural project to fully integrate these principles, a project has to be viewed holistically; which means considering all aspects of a project—social, economic and environmental. A useful starting point in this regard is the value chain approach. A value chain "describes the full range of activities required to bring a product or service from conception, through the different phases of production (involving a combination of physical transformation and the input of various producer services), delivery to final consumers and final disposal after use" (Kaplinsky and Morris 2001). A value chain analysis is useful for identifying and understanding the nature of and extent of activities associated with a project, and provides opportunities for identifying areas where socio-economic principles can be realistically integrated.

However, in many cases, those involved in one stage of a value chain have virtually no control over what happens in other stages. For example, a farmer producing a crop generally has no control over the way inputs are produced and distributed, nor the processing of the crop after it is sold. In such situations, efforts to integrate socio-economic green economy principles into a project should be focused on those aspects which are within the control of those working on agricultural projects. This should not, however, dissuade those implementing green economy projects from adopting a value chain approach, but illustrates the multidimensional nature of the green economy as it relates to agriculture, and the need to take a flexible approach to integrating socio-economic principles into a project.

A green economy project does not occur in a vacuum. The integration of socio-economic principles has to occur within a project's context. As highlighted in Sect. 2.6, in South Africa, for example, addressing poverty and inequality are national priorities articulated in various policies and strategies. A green economy project in South Africa has to strive to contribute to social equity and inclusivity, and to addressing poverty. A project could do this for example by employing vulnerable groups, especially the poor, women, youth and differently abled people. A project would also have to strive to create jobs and decent work; e.g. through the choices that are made throughout the project value chain. For example, minimizing mechanisation and using labour-based methods is one way of maximising job creation. At the same time, harnessing new technology and digital opportunities may open up new green avenues and modes of production.

The local socio-economic context has to inform and be integrated into project implementation. In South Africa, rural development and Local Economic Development (LED) are highlighted as priorities in government development plans, for example the Medium Term Strategic Framework (Presidency Republic of South

Africa 2009) and the National Strategy for Sustainable Development and Action Plan (DEA 2011). A green economy project should therefore not be inward looking, but should endeavour to contribute to the general development of the local area and to the livelihoods and well-being of local people. Given the limited economic opportunities in rural areas and the fact that agricultural projects have a high likelihood of being located in rural areas, contribution to rural and local economic development is therefore a key aspect of integrating socio-economic green economy principles. Green economy projects have to contribute to addressing the needs of communities in the areas in which they are located. Projects could, for example, employ local people and sell some produce to local consumers and traders, thus contributing to local food security.

Finally, a requirement for any successful agricultural project is its ability to meet its objectives, such as food production on a sustained basis, and to be economically viable. Factors such as crop choice and suitability for local conditions, production practices and marketing of produce all have a bearing on a project's sustainability and viability. A green economy project should be set up and implemented in a way that optimises these factors so as to ensure profitability and sustainability. This is important, as it underpins the well-being of farmers and others who intend to derive a livelihood from such a project. Practices such as sustainable agriculture, which aims to use farming methods that are economically profitable while protecting the environment, human health and communities (Kirchmann and Thorvaldsson 2000; Pretty et al. 2008); are well aligned with socio-economic green economy principles; and present opportunities for translating these ideals into tangible benefits for people.

2.10 Conclusion

The green economy is a complex, multidimensional concept which aims to simultaneously address social, economic, and environmental issues, and to achieve multiple objectives. Being a multi-faceted concept, the green economy has to address issues from different perspectives, and this entails utilizing learning and information from diverse sources. Although the term 'green economy' is often interpreted in a narrow environmental sense; many of the principles of a green economy do in fact refer to socio-economic imperatives. These include, for example, principles related to social inclusivity, poverty reduction, well-being and livelihoods, decent work and job creation. Nevertheless, some of these principles may seem abstract and far removed from agriculture at the practical field level. This chapter identifies and discusses the factors that should inform green economy implementation in an agricultural context to ensure that socio-economic issues are adequately addressed and different objectives are achieved. These factors, which include green economy principles; local and national development issues, global trade factors, human rights issues, disruptive technologies and commodification of natural resources are analysed in a developing country context. This chapter has highlighted the variety of issues necessary to consider from a socio economic perspective. In Chap. 3 we examine the biophysical and environmental context of green economy implementation.

References

Aceleanu MI, Serban AC, Burghelea C (2015) Greening the youth employment — a chance for sustainable development. Sustainability 7:2623–2643

Adeleke S, Kamara AB, Zuzana B (2010) Smallholder agriculture in East Africa: trends, constraints and opportunities, working papers series No. 105 African Development Bank, Tunis, Tunisia. https://www.commdev.org/wp-content/uploads/2015/06/Smallholder-Agriculture-East-Africa-Trends-Constraints-Opportunities.pdf. Accessed 02 Sept 2018

Allen C (2012) A guidebook to the green economy. Issue 2: exploring green economy principles. United Nations Department of Economic and Social Affairs (UNDESA): united nations division for sustainable development

Africa Research Institute (2013) Waiting for the green revolution: land reform in South Africa. http://www.africaresearchinstitute.org/newsite/wp-content/uploads/2013/05/BN1301-South-Africa-Land-Reform1.pdf. Accessed 27 Sept 2018

AFRICEGE (2015) Mapping the green economy landscape in South Africa. http://www.sagreenfund.org.za/wordpress/wp-content/uploads/2015/04/Mapping-the-Green-Economy-in-SA.pdf. Accessed 13 March 2018

AGRA (2017) Africa agriculture status report: the business of smallholder agriculture in sub-Saharan Africa (Issue 5). Nairobi, Kenya: alliance for a green revolution in Africa (AGRA). Issue No. 5. https://agra.org/wp-content/uploads/2017/09/Final-AASR-2017-Aug-28.pdf. Accessed 02 Sept 2018

Aliber A, Hart T (2009) Should subsistence agriculture be supported as a strategy to address rural food insecurity? Agrekon 48:434–458

African Union (2011) Assembly of the union: seventeenth ordinary session: decisions, declarations and resolution. 30 June - 1 July 2011; Malabo, Equatorial Guinea. https://au.int/sites/default/files/decisions/9647-assembly_au_dec_363-390_xvii_e.pdf. Accessed 09 Oct 2018

ASFG (2010) Africa's smallholder farmers: approaches that work for viable livelihoods. A report by the African smallholder farmers group (ASFG). http://www.asfg.org.uk/downloads/final-asfg—africas-smallholder-farmers.pdf. Accessed 02 Sept 2018

AU and NEPAD (2018) Drones on the horizon: transforming Africa's agriculture. http://www.nepad.org/resource/drones-horizon-transforming-africas-agriculture. Accessed 18 June 2018

Babugura A (2017) Gender equality: a cornerstone for a green economy. South African institute of international Affairs. Occasional Paper 269. http://www.saiia.org.za/research/gender-equality-a-cornerstone-for-a-green-economy/. Accessed 20 Aug 2018

Banerjee AV, Duflo E (2011) Poor economics: a radical rethinking of the way to fight global poverty. Public Affairs, New York

Bergius M, Benjaminsen TA, Widgren M (2018) Green economy, Scandinavian investments and agricultural modernization in Tanzania. J Peasant Stud 45:825–852

Berkey RE (2014). Just farming: an environmental justice perspective on the capacity of grassroots organisations to support the rights of organic farmers and laborers. Doctor of philosophy. Environmental studies. Antioch University New England

Bradley K, Herrera H (2016) Decolonising food justice: naming, resisting, and researching colonizing forces in the movement. Antipode 48(1):97–114

Bryant B (1995). Issues and potential policies and solutions for environmental justice: an overview. Environmental justice: issues, policies, and solutions, pp 8–34

Burnett K, Murphy S (2014) What place for international trade in food sovereignty? J Peasant Stud 41:1065–1084

Buseth JT (2017) The green economy in Tanzania: from global discourses to institutionalization. Geoforum 86:42–52

Christensen CM (1997) The innovator's dilemma: when new technologies cause great firms to fail. Harvard Business Review Press, Boston

City of Tshwane (2013) Framework for a green economy transition. http://resilientcities2015. iclei.org/fileadmin/RC2015/files/Framework_for_a_Green_Economy_Transition.pdf. Accessed 24 Sept 2018

Clark M, Luwaya N (2017) Communal land tenure 1994–2017: commissioned report for high level panel on the assessment of key legislation and the acceleration of fundamental change, an initiative of the parliament of South Africa. https://www.parliament.gov.za/storage/app/media/Pages/2017/october/High_Level_Panel/Commissioned_Report_land/Commisioned_Report_on_Tenure_Reform_LARC.pdf. Accessed 15 Aug 2018

DEA (2011). National strategy for sustainable development and action plan (NSSD 1) 2011–2014

DEA (2017) South Africa's green economy strategy. http://www.enviropaedia.com/topic/default.php?topic_id=342. Accessed 7 Sept 2018

DEAT (2008) People-planet-prosperity. A national framework for sustainable development in South Africa. https://www.environment.gov.za/sites/default/files/docs/2008nationalframeworkfor_sustainabledevelopment.pdf. Accessed 20 Sept 2018

DEDT KwaZulu Natal (2012). Department of economic development and Tourism KwaZulu-Natal Province. Developing a strategy for a green economy in KwaZulu-Natal. Volume 4: green economy strategy for KwaZulu-Natal Province. http://www.kznded.gov.za/Portals/0/DEDTGreenEconStrategy_DraftStrategy30March2012%20_2_.pdf. Accessed 12 Aug 2018

EDD (2011a) The new growth path: framework. http://www.economic.gov.za/communications/publications/new-growth-path-series. Accessed 20 Aug 2018

EDD (2011b) New growth path: accord 4 green economy accord. http://www.economic.gov.za/communications/publications/green-economy-accord. Accessed 4 Oct 2018

Ederera S, Heumesserb C, Staritz C (2016) Financialisation and commodity prices – an empirical analysis for coffee, cotton, wheat and oil. Int Rev Appl Econ 30:462–487

Elliot KE (2018) The WTO, agriculture, and development: a lost cause? https://www.ictsd.org/bridges-news/bridges-africa/news/the-wto-agriculture-and-development-a-lost-cause. Accessed 18 June 2018

Ehresman TG, Okereke C (2015) Int Environ Agreem 15:13–27. https://doi.org/10.1007/s10784-014-9265-2

FAO (2003) World agriculture: towards 2015/2030 an FAO perspective. http://www.fao.org/3/a-y4252e.pdf. Accessed 16 July 2018

FAO and ITU (2018) E-agriculture in action: drones for agriculture. http://www.fao.org/3/i8494en/I8494EN.pdf. Accessed 16 July 2018

Gabara N (2012) Small-scale farmers encouraged to drive economy. South African government news agency. http://www.sanews.gov.za/business/small-scale-farmers-encouraged-drive-economy. Accessed 20 Dec 2014

Gauteng Province (2011) Department of economic development, Gauteng provincial government green strategic programme for Gauteng. http://www.ecodev.gpg.gov.za/policies/Documents/Gauteng%20Green%20Strategic%20Programme.pdf. Accessed 30 June 2014

Gollin D (2014) Smallholder agriculture in Africa: an overview and implications for policy IIED working paper. IIED, London. http://pubs.iied.org/pdfs/14640IIED.pdf. Accessed 2 Sept 2018

Goodman MK (2004) Reading fair trade: political ecological imaginary and the moral economy of fair trade foods. Polit Geogr 23:891–915

Gottlieb R, Fisher A (1996) Community food security and environmental justice: searching for a common discourse. Agric Hum Values 13:23–32

Green Economy Coalition (2012) A national vision for a green economy emerges from Botswana. https://www.greeneconomycoalition.org/news-analysis/national-vision-green-economy-emerges-botswana. Accessed 18 Sept 2018

Guillén-Navarro MÁ, Pereñíguez-García F, Martínez-España R (2017) IoT-based system to forecast crop frost. In: 13th international conference on intelligent environments, pp 28–35. http://agri. ckcest.cn/ass/af563fa2-b946-4e97-a2d8-bfd7cf17a13a.pdf. Accessed 16 July 2018

Haaland C, van den Bosch CK (2015) Challenges and strategies for urban green-space planning in cities undergoing densification: a review. Urban Urban Green 14(4):760–771

Hall JK, Martin MJC (2005) Disruptive technologies, stakeholders and the innovation value-added chain: a framework for evaluating radical technology development. R&D Manag 35:273–284

Halle M (2011) Accountability in the green economy. In: beyond Rio + 20: governance for a green economy. Pardee Center Task Force Report/ March 2011. http://www.bu.edu/pardee/files/2011/ 03/Rio20TFC-Mar2011.pdf. Accessed 15 Aug 2018

ILO (2016) What is a green job? http://www.ilo.org/global/topics/green-jobs/news/WCMS_ 220248/lang–en/index.htm. Accessed 20 Aug 2018

ILO (2018) World employment social outlook: trends 2018. ILO, Geneva. https://www.ilo.org/ wcmsp5/groups/public/—dgreports/—dcomm/—publ/documents/publication/wcms_615594. pdf. Accessed 30 July 2018

ILO, UNCTAD (2013) Shared harvests: agriculture, trade and employment. International labour office and united nations conference on trade development - Geneva. http://unctad.org/en/ PublicationsLibrary/ditctncd2013d2_en.pdf. Accessed 24 Sept 2018

IFAD (2015) Land tenure security: scaling up note. https://www.ifad.org/documents/38714170/ 40196966/Scaling+up+results+in+land+tenure+security.pdf/9be8e8e7-1a76-4b2c-9ab6- 328f6c20df67. Accessed 15 Sept 2018

Internet World Stats (2018) Internet users statistics for Africa. https://www.internetworldstats.com/ stats1.htm. Accessed 24 Sept 2018

Kaplinsky R, Morris M (2001) A handbook for value chain research. https://www.ids.ac.uk/ids/ global/pdfs/ValuechainHBRKMMNov2001.pdf. Accessed 24 Sept 2018

Karapinar B (2010) Introduction: food crises and the WTO. In: Karapinar B, Häberli C (eds) Food crises and the WTO: world trade forum. Cambridge University Press, Cambridge

King A (2017) The future of agriculture. Nature 544(7651):S21–S23. https://www.nature.com/ articles/544S21a.pdf. Accessed 24 Sept 2018

Kingdon G, Knight J (2004) Unemployment in South Africa: the nature of the beast. World Dev 32(3):391–408

Kirchmann H, Thorvaldsson G (2000) Challenging targets for future agriculture. Eur J Agron 12:145–161

Krippner G (2011) Capitalising on crisis: the political origins of the rise of finance. Harvard University Press, Cambridge, MA

Levidow L (2014) What green economy? Diverse agendas, their tensions and potential futures. IKD working paper No. 73. https://oro.open.ac.uk/40808/1/LL_What%20Green%20Economy_ IKD%20WP_2014.pdf. Accessed 30 July 2018

Lutz C, Olthaar M (2017) Global value chains and smallholders in sub-Saharan Africa. Rev Soc Econ 75:251–254

Marston A (2013) Justice for all? Material and semiotic impacts of fair trade craft certification. Geoforum 44:162–169

Moodley S (2013) By declaring 2014 the 'year of agriculture', the African Union hopes to spur a green revolution. http://www.engineeringnews.co.za/article/by-declaring-2014-the- year-of-agriculture-the-au-hopes-to-spur-a-green-revolution-2013-09-27/rep_id:4136. Accessed 09 Oct 2018

Mudombi S (2017) Using the green economy and youth inclusion for sustainable development in South Africa. In: Trade and industrial policy studies POLICY BRIEF 8/2017 October. http://www.tips.org.za/policy-briefs/item/3395-using-the-green-economy-and- youth-inclusion-for-sustainable-development-in-south-africa. Accessed 20 Aug 2018

NDA (1998) Agricultural policy in South Africa: a discussion document. Ministry for agriculture and land affairs. http://www.nda.agric.za/docs/Policy/policy98.htm. Accessed 14 April 2014

NDA (2002) The strategic plan for South African agriculture. http://www.nda.agric.za/docs/sectorplan/Vouer_e.htm. Accessed 11 April 2014

NDA (2003) Draft sustainable utilisation of agricultural resources bill. http://www.daff.gov.za/docs/bills/sustainable.htm. Accessed 11 April 2014

Newman S (2009) Financialisation and changes in the social relations along commodity chains: the case of coffee. Rev Radic Polit Econ 41(4):539–559

Nhamo G, Chekwoti C (2014) New generation land grabs in a green African economy. In: Nhamo G, Chekwoti C (eds) Land grabs in a green African economy. Implications for trade, investment and development policies. Africa Institute of South Africa, Pretoria, pp 1–9

Nicholls S, Vermaak M, Moolla Z (2016) The power of collective action in green economy planning. It's the economy, stupid. The National Business Initiative. Green Fund, Development Bank of Southern Africa, Midrand

NPC (2011) National development plan, 2030. Our future make it work. Republic of South Africa

Otte J (2007) Globalisation and smallholder farmers. A pro-poor livestock policy initiative. Res Rep. http://www.fao.org/3/a-bp293e.pdf. Accessed 26 Sept 2018

Ouma S (2014) Situating global finance in the land rush debate: a critical review. Geoforum 57:162–166

OXFAM (2010) A fresh look at the green economy: Jobs that build resilience to climate change. https://www.oxfamamerica.org/static/media/files/a-fresh-look-at-the-green-economy.pdf. Accessed 30 July 2018

PAGE (2017), Green economy inventory for South Africa: an overview. Pretoria. South Africa. http://thegreentimes.co.za/wp-content/uploads/2017/08/green_economy_inventory_for_south_africa.pdf. Accessed 15 July 2018

Pellow DN (2000) Environmental inequality formation: toward a theory of environmental injustice. Am Behav Sci 43:581–601

Peralta A (2017) From the financialisation of food to life-enhancing agriculture. In: Peralta A (ed) Food and Finance: toward life-enhancing agriculture. World Council of Churches, Geneva, Switzerland. https://www.oikoumene.org/en/resources/publications/TheFinancializationofFood.pdf. Accessed 24 August 2018

Pettinger T (2018) Neoliberalism – examples and criticisms. https://www.economicshelp.org/blog/20688/concepts/neoliberalism/. Accessed 26 Sept 2018

Piñeiro V, Piñeiro M (2017). The future of the global agri-food trade and the WTO. In: Piñeiro V, Piñeiro M (eds) Agricultural trade interests and challenges at the WTO ministerial conference in Buenos Aires: A Southern Cone perspective. http://ebrary.ifpri.org/cdm/ref/collection/p15738coll2/id/131544. Accessed 18 Aug 2018

Pirnea IC, Lanfranchi M, Giannetto C (2013) Agricultural market crisis and globalisation – a tool for small farms. Revista Românã de Statisticã 10:35–45

Presidency Republic of South Africa (2009) Together doing more and better: medium term strategic framework. A framework to guide government's programme in the electoral mandate period (2009–2014)

Pretty J, Smith G, Goulding KWT, Groves SJ, Henderson I, Hine RE, King V, van Oostrum J, Pendlington DJ, Vis JK, Wlater C (2008) Multi-year assessment of Unilever's progress towards agricultural sustainability I: indicators, methodology and pilot farm results. Int J Agric Sustain 6:37–62

Scotcher JSB (2009) The green choice living farms reference 2009/2010 version. In: Goldblatt A (ed) Unpublished report to Green Choice (a WWF and Conservation International partnership)

Stafford W, Facer K, Audouin M, Funke N, Godfrey L, Haywood L, Musvoto C, Strijdom W (2014) Steering towards a green economy: a reference guide. https://www.csir.co.za/sites/default/files/Documents/GE%20guide.pdf. Accessed 20 Sept 2018

Staritz C, Newman S, Tröster B, Plank L (2015) Financialisation, price risks, and global commodity chains: distributional implications on cotton sectors in sub-Saharan Africa. Austrian foundation for development research – ÖFSE. http://www.oefse.at/fileadmin/content/Downloads/Publikationen/Workingpaper/WP55_Financialization.pdf. Accessed 23 Aug 2018

StatsSA (2013) Gender statistics in South Africa, 2011. http://www.statssa.gov.za/publications/Report-03-10-05/Report-03-10-052011.pdf. Accessed 20 Aug 2018

StatsSA (2017a) StatsSA quarter four labour force survey. http://www.statssa.gov.za/publications/P0211/P02111stQuarter2018.pdf. Accessed 20 Aug 2018

StatsSA (2017b) Poverty trends in South Africa: an examination of absolute poverty between 2006 and 2015. https://www.statssa.gov.za/publications/Report-03-10-06/Report-03-10-062015.pdf. Accessed 20 Aug 2018

StatsSA (2018a) Youth unemployment still high in Q1: 2018. http://www.statssa.gov.za/?p=11129. Accessed 20 Aug 2018

StatsSA (2018b) Quarterly labour force survey, quarter 4, 2017. http://www.statssa.gov.za/publications/P0211/P02114thQuarter2017.pdf. Accessed 20 Aug 2018

Sustainable Development Solutions Network (2014) Why good governance of land and tenure security need to be part of the sustainable development goal framework. www.focusonland.com/download/547dff4f70e29/. Accessed 27 Sept 2018

Swilling M, Kaviti Musango J, Wakeford J (2016) Introduction: deepening the green economy discourse in South Africa. In Swilling M, Kaviti Musango J, Wakeford J (eds) Greening the South African economy; scoping the issues, challenges and opportunities, 1st edn. UCT Press

Swinnen J, Colen L, Maertens M (2013) Constraints to smallholder participation in high-value agriculture in West Africa, In: Elbehri A (ed) Rebuilding West Africa's food potential, FAO/IFAD. http://www.fao.org/docrep/018/i3222e/i3222e09.pdf. Accessed 7 June 2018

Sze J, London JK (2008) Environmental justice at the crossroads. Sociol Compass 2:1331–1354

TRALAC (2016) Promoting agricultural global (regional) value chains in Africa. https://www.tralac.org/discussions/article/10923-promoting-agricultural-global-regional-value-chains-in-africa.html. Accessed 7 June 2018

UNDP (2014) Environmental Justice: comparative experiences in legal empowerment. http://www.undp.org/content/undp/en/home/librarypage/democratic-governance/access_to_justiceandruleoflaw/environmental-justice—comparative-experiences.html. Accessed 28 Sept 2018

Unmüßig B, Sachs B, Fatheuer T (2012) Critique of the green economy: toward social and environmental equity. Edited by the Heinrich Böll Foundation. https://us.boell.org/sites/default/files/downloads/Critique_of_the_Green_Economy.pdf. Accessed 23 Aug 2018

UNCTAD (2009) World Investment Report 2009. Transnational Corporations, Agricultural Production and Development. https://unctad.org/en/docs/wir2009_en.pdf. Accessed 26 Sept 2018

UNEP (2011) Green economy: why a green economy matters for the least developed countries. http://unctad.org/en/docs/unep_unctad_un-ohrlls_en.pdf. Accessed 20 Aug 2018

UNEP (2012) Principles for a green, fair and inclusive economy Version 3. http://www.unep.org/greeneconomy/Portals/88/documents/GEI%20Highlights/Principles%20of%20a%20green%20economy.pdf. Accessed 15 Sept 2013

UNEP (2013) Green economy scoping study: South African green economy modelling report (SAGEM) – Focus on natural resource management, agriculture, transport and energy sectors. http://wedocs.unep.org/bitstream/handle/20.500.11822/18316/SAModellingReport.pdf?sequence=1&isAllowed=y. Accessed 05 Dec 2018

Vermeire JAL, Bruton GD, Cai L (2017) Global value chains in Africa and development of opportunities by poor landholders. Rev Soc Econ 75:280–295

von Braun J, Meinzen-Dick R (2009) 'Land grabbing' by foreign investors in developing countries: risks and opportunities. IFPRI policy brief, vol 13. International Food Policy Research Institute, Washington, DC

WEF (2017) Shaping the future of global food systems: a scenarios analysis. http://www3.weforum.org/docs/IP/2016/NVA/WEF_FSA_FutureofGlobalFoodSystems.pdf. Accessed 19 Aug 2018

WEF (2018) Innovation with a purpose: the role of technology innovation in accelerating food systems transformation http://www3.weforum.org/docs/WEF_Innovation_with_a_Purpose_VF-reduced.pdf. Accessed 19 Aug 2018

Western Cape Government (2013). Green is smart. Western cape green economy strategy frame-
    work. https://www.westerncape.gov.za/assets/departments/transport-public-works/Documents/
    green_is_smart-4th_july_2013_for_web.pdf. Accessed 24 Sept 2018
WHL (2018) Sustainable farming and sourcing. https://www.woolworthsholdings.co.za/
    sustainable/. Accessed 24 Aug 2018
Wolch JR, Byrne J, Newell JP (2014) Urban green space, public health, and environmental justice:
    the challenge of making cities 'just green enough'. Landsc Urban Plan 125:234–244
World Bank (2016a) Poverty and shared prosperity 2016: taking on inequality. https://doi.org/10.
    1596/978-1-4648-0958-3
World Bank (2016b) World development report 2016: digital dividends. http://www.worldbank.org/
    en/publication/wdr2016
World Bank (2017) Agriculture and food. http://www.worldbank.org/en/topic/agriculture/overview.
    Accessed 30 July 2018

Chapter 3
The Biophysical and Environmental Context

3.1 Biophysical Issues in a Crop Production Context

Crop production relies on natural capital[1] which is made up of the biophysical environment. Climatic conditions are an important component of the biophysical environment. Each crop has a particular optimum in terms of growing temperature, humidity, day-length, soil moisture, soil type and nutrient requirements. When these conditions are not optimal, growth and yield are reduced. Managing the biophysical environment to maintain optimum production is an important aspect of agriculture. Some of the biophysical conditions which influence crop growth are not easy to alter, while others can easily be altered through agricultural practices and management. In a green economy context, production has to be optimum and the practices used should not undermine natural capital, but have to contribute to its maintenance and enhancement. Some of the key biophysical issues that green economy projects should be cognisant of and which implementation should address are discussed below.

Temperature is one of the major environmental factors affecting the growth, development and yields of crops (Luo 2011). Growth rates and yields are maximised when crops are grown at their optimal temperature, and gradually decrease at lower temperatures until no development occurs; similarly, at temperatures higher than the optimum, development rates decline until plant death occurs (Porter and Gawith 1999). The earth's thermal zones (Köppen 2011) provide a broad indication of the expected temperature and climatic conditions in an area. However, local variations occur as factors such as elevation and aspect affect local climates. Since specific crops are suited to particular conditions, yields are influenced by climate, although management factors such as levels of inputs (e.g. fertilisers) also affect yields. Green economy projects have to ensure that crops are matched with appropriate climatic conditions to ensure maximum productivity.

---

[1] Natural capital can be defined as the world's stocks of natural assets which include geology, soil, air, water and all living things (Natural Capital Forum 2018).

Water is essential for crop production; and an adequate amount of water is vital for plant growth and maintenance of essential processes. Crops have different water needs, as some crops use water more efficiently than others (Gurian Sherman 2012). Crops can meet their water needs from rainfall and/or irrigation. Water stress decreases the water potential of plants, inhibits photosynthesis, and reduces growth and yield (Kirkham 1990; Gupta et al. 2001; Ontel and Vladut 2015). Crop production in many parts of the world is seriously limited by lack of water, and water supplies for agriculture are also dwindling (Morison et al. 2008). At the other extreme water constraining plant growth is too much water, which also constrains plant growth due to poor drainage and waterlogging. Poor drainage reduces the space for oxygen in the rooting zone and this could lead to oxygen deprivation to plants and this negatively affects growth and yield. However, crops differ in their tolerance to waterlogging. In addition to oxygen reduction in the rooting zone, waterlogging increases the incidence and severity of soil-borne pathogens and also makes land access and tillage difficult (Laidlaw 2009; Keane 2001; Jones and Thomasson 1993).

In cropping systems water should be managed to ensure that crop water requirements are met. Under rainfed conditions, crop water requirements should match the rainfall of the area in which the crop is planted. Irrigation allows crops to be grown in areas where rainfall is not enough to meet crop water requirements. When irrigation is used, it is important to apply efficient water and crop management strategies to optimise the use of water. These strategies include optimal choice of irrigation system, proper irrigation scheduling in terms of both timing and quantity of water applied and best crop management with regards to the soil and climate conditions (Mancosu et al. 2015). Green economy projects should follow best practices to ensure adequate water provision to crops.

Soil is one of the most critical resources for crop production as it is the natural medium for plant growth. Soil is comprised of minerals, soil organic matter (SOM), water, and air (FAO 1987). Soil components influence soil physical properties, including texture, structure, and porosity, the fraction of pore space in a soil; and these in turn affect air and water movement in the soil, and thus the soil's ability to function (McCauley et al. 2005). Soil also hosts a complex web of fauna, flora and microorganisms, and these are involved in many different biological processes, which also affects a soil's physical and chemical properties, and ultimately the productivity of agricultural ecosystems (McCauley et al. 2005; Delgado and Gómez 2016).

Soil directly and indirectly affects agricultural productivity, water quality, and the global climate through its function as a medium for plant growth, and as regulator of water flow and nutrient cycling (Delgado and Gómez 2016). Soils are the natural source of nutrients required for crop growth. These nutrients can also be provided through fertilisation. However, the over-application of fertilisers, particularly chemical fertilisers that readily leach from soils is detrimental to soil health and to the environment. It is important to manage soils to ensure that they are able to sustain crop production. Soil degradation, defined in this context as the diminished capacity of a soil to perform selected specific services such as growing crops (Hatfield et al. 2017) is a major problem for agriculture. Degradation of soil results in the loss of critical functions and ecosystem services which include crop production, ensuring sufficient supplies of clean water, acting as a buffer against extreme climatic events, supporting biodiversity, and providing the largest terrestrial store of carbon and nutrients (Janzen et al. 2011). Soil degradation, therefore, poses a threat to food security, as it reduces yield, forces farmers to use more inputs, and may eventually lead to land abandonment (Gomiero 2016). Although many factors cause soil degradation, agriculture is a dominant issue with factors such as excessive tillage, inappropriate crop rotations and crop residue removal causing degradation (Karlen and Rice 2015). The role of agriculture in soil degradation is such that green economy implementation should take special care to protect soil. Enhancement and maintenance of soil productivity is essential for the sustainability of agriculture (Lal and Stewart 1995).

Day length (the duration of the light period), also referred to as photoperiod is an important biophysical factor in crop production as it affects crop growth and development. The latitude of a site determines day length. In addition to photoperiod, Crauford and Wheeler (2009) note that site latitude is associated with variations in temperature regimes and radiation intensity, all of which determine crop growth and development and ultimately productivity. As each crop has specific day length requirements, green economy projects should ensure that crops are correctly matched with sites to ensure maximum productivity.

One of the important biophysical factors in crop production is agricultural biodiversity or agrobiodiversity. Agricultural biodiversity refers to the variety and variability of animals, plants, and micro-organisms on earth that are important to food and agriculture (FAO 2004). Agricultural biodiversity also encompasses the diversity of crops and their wild relatives, trees, livestock and landscapes (Tutwiler et al. 2017) and is critical for complexity of agricultural systems. The loss of agricultural diversity is a growing global concern (Marvier 2001) as it leads to the simplification of farming systems. Simple farming systems leave farmers with a decreasing range of resources to draw on to manage threats such as pests and diseases, declining soil fertility, or the impacts associated with increasing climatic variability (Attwood et al. 2017). In order to address these and many other issues, sustainable practices are needed and agricultural biodiversity is a key component of sustainable practices (Attwood et al. 2017). Agricultural biodiversity is a critical component of sustainable food systems (Tutwiler et al. 2017) and agricultural systems in general. Agricultural biodiversity provides the diversity which helps to drive critical ecological processes (e.g. soil structure maintenance) and allows the simultaneous provision of multiple benefits (including nutritious foods, income, natural pest control, pollination, water quality) (Attwood et al. 2017). Agricultural biodiversity-based strategies are noted to be important for soil erosion control, climate resilience, pest and disease control, productivity, pollination and wild biodiversity conservation (Attwood et al. 2017). Green economy implementation should be cognisant of the importance of maintaining biodiversity and should take special care to incorporate agricultural biodiversity-based strategies.

3.2 The Environmental Impacts of Agriculture—Implications for the Green Economy and Its Implementation

Agriculture and nature are inextricable, with agriculture being dependent on a healthy natural environment for the ecosystem services which underpin agricultural productivity. These ecosystem services are essential for good quality and healthy food, and underpin the development of strong rural economies and local communities (WWF 2017). At the same time, agricultural activities necessitate the transformation of the natural environment into agroecosystems. This transformation impacts on the environment in various ways, including deforestation, loss of biodiversity and soil erosion. Other environmental impacts relate to physical, biological and chemical degradation of soils and water. According to World Wildlife Fund (WWF) (2017), unsustainable farming practices are an important driver of biodiversity loss and environmental degradation; and when practised without care, farming presents the greatest threat to species and ecosystems.

The New Partnership for Africa's Development (NEPAD) (2002) noted with concern the environmental degradation caused by agriculture in many parts of Africa. NEPAD indicated that, in many places, environmental degradation and unsustainable exploitation of natural resources threaten to reduce the future productivity of agriculture and natural resources, and that a major challenge for African countries was to ensure that agriculture does not degrade the underlying natural resource base (NEPAD 2002). Similar concerns have been expressed by the Millennium Ecosystem Assessment (MEA). According to the MEA (2006), two-thirds of the Earth's ecosystem services are in decline while the resources humans depend on for much of the world's food supply are finite, declining, and in some cases, disappearing. Freshwater is becoming scarcer, land is degraded and ecosystems are in decline (MEA 2006). Farming practices which degrade the environment are not only detrimental to the environment, they also undermine agriculture. According to IFAD (2013), current practices are undermining the ecological foundation of the global food system through overuse and the effects of agricultural pollution, thereby enhancing degradation and reducing ecosystem capacity to generate sustainable yields. The loss of agricultural productivity also negatively impacts on the backward and forward linkages of agriculture with other sectors of the economy. For example, it will result in a shortage of raw materials inputs for the manufacturing sector (Scotcher 2009).

According to UNEP (2012), more than 20% of cultivated lands have decreasing productivity due to degradation; with much of the degrading land being in Africa south of the equator (13% of the global degrading area). There are agricultural practices that pose particular risks to the environment, including monocultures and tillage. Monocultures reduce on-farm biodiversity (which decreases the resilience of crops to pests and diseases), ecosystem functions and ecological resilience, and increase vulnerability to environmental risks, such as climate change (IAASTD 2009; UNEP 2012). Tillage, defined as the mechanical manipulation of the soil for the purpose of crop production, significantly affects the physical, chemical and biological properties of soil; and characteristics such as soil water conservation, soil temperature, infiltration and evapotranspiration (Busari et al. 2015). Excessive tillage disrupts natural soil structure and promotes soil loss and decline of overall soil quality (Karlen et al. 2013). Furthermore, injudicious use of inputs such as inorganic fertilisers and pesticides can have negative impacts on the environment. The impacts of high fertiliser loading on the environment include eutrophication of surface waters and contamination of groundwater (UNEP 2012; IAASTD 2009). In many farming systems, the excessive use of irrigation, pesticides and fertilisers has been a major cause of impacts such as soil acidification and salinization, eutrophication and contamination (Tilman et al. 2002; Cassman et al. 2003; Hochman et al. 2013).

The environmental impacts of agriculture are not limited to large-scale conventional[2] operations. Small-scaSle farming, although generally utilising low levels of inputs and little machinery, can also have adverse effects on the environment. For example, in South Africa, soil degradation has been reported to be most severe in communal croplands and grazing lands (Meadows and Hoffmann 2002) where small-scale agriculture is practised. IFAD (2013) notes that smallholder farming affects the condition of ecosystems; and the need to satisfy immediate needs can drive smallholders to adopt environmentally damaging agricultural practices, resulting in soil erosion, nutrient depletion, salinization, water scarcity and pollution.

Agriculture is one of the largest contributors to anthropogenic climate change through greenhouse gas emissions. Emissions from agriculture are varied, and include carbon dioxide, nitrous oxide and methane. Global food systems are responsible for 19–29% of all anthropogenic greenhouse gas (GHG) emissions (Vermeulen et al. 2012). Agricultural expansion through habitat conversion accounts for about 70% of land use change emissions, mainly through deforestation (Hosonuma et al. 2012; IPCC 2014a, b; Tubiello et al. 2015), and is responsible for biodiversity loss (Lanz et al. 2017) and its associated impacts on ecosystem services.

The green economy espouses low carbon development, conservation of natural resources, and minimising damage to the environment; whilst still meeting human needs. Practising agriculture in a way that impacts negatively on the environment is thus at odds with the green economy. Green economy implementation in the agriculture sector has to address this conflict. Agricultural green economy projects therefore have to not only focus on meeting production objectives, but on also reducing their environmental impact and providing ecological services. If farming operations are sustainably managed, they can help preserve and restore the environment and the critical services it provides. UNEP (2012) describe agriculture in a green economy context as one that 'involves the application of food production and consumption practices that ensure productivity and profitability without undermining ecosystem services, and rebuilds ecological resources by reducing pollution and using resources more efficiently'.

---

[2]Conventional farming is also known as industrial agriculture and refers to resource and energy intensive farming systems which include the use of inputs such as synthetic chemical fertilisers, pesticides, herbicides, genetically modified organisms, heavy irrigation, intensive tillage, or concentrated monoculture production (http://www.appropedia.org/Conventional_farming).

3.3   Environmental Issues in Relation to Green Economy Principles and Implications for Agricultural Practices

Historically, the agriculture sector has sought to maximise production and minimise costs; in many instances with little regard to its impacts on the environment or to its role in society. However, the increasing global focus on sustainable development has seen an expanded awareness of, and movement towards also improving sustainability in agriculture. In particular, there has been a focus on techniques and practices aimed at reducing the environmental impacts of agriculture. More recently, with the green economy concept being adopted into the mainstream, there has been further recognition that agriculture will have to be practised in such a way that it adheres to the principles of a green economy.

Given the environmental problems that beset the agriculture sector (see Sect. 3.2), for agriculture to support a green economy, the negative environmental impacts of the sector have to be addressed. While green agriculture is not synonymous with a green economy, agriculture has to be 'green' if it is to contribute towards a green economy. According to UNEP (2011: 42), "the greening of agriculture refers to the increasing use of farming practices and technologies that simultaneously:

- maintain and increase farm productivity and profitability while ensuring the provision of food and ecosystem services on a sustainable basis;
- reduce negative externalities and gradually lead to positive ones; and
- rebuild ecological resources (i.e. soil, water, air and biodiversity natural capital assets) by reducing pollution and using resources more efficiently."

In addition, in Chap. 1, we identified 11 green economy principles, based on an assessment conducted by UNDESA (Allen 2012). A number of these principles speak directly to environmental issues, and have clear implications for agricultural practices that should be applied in green economy implementation.

In particular, the following implications for agriculture are clear

- Agricultural projects should strive to be resource and energy efficient
- Agricultural projects should strive to be 'low carbon'
- Agricultural projects must support environmental protection, and in particular the protection of biodiversity and ecosystems.

In short, agriculture will have to be practised in such a way that it respects planetary boundaries, ecological limits and scarcities; and does not undermine the integrity of the environmental systems on which it depends. In the following sub-sections, we briefly discuss each of these implications. Section 3.5 goes into more detail regarding specific agricultural practices and methods that can be adopted in line with green economy principles.

3.3.1 Resource and Energy Efficiency in Agriculture

In an increasingly resource-constrained world, and given the growing demand for agricultural output to meet the food needs of an expanding population, it is imperative that agriculture operates in a resource efficient way. These resources include land, water, energy, and other production inputs, such as fertilisers. For example, there is a critical need to switch toward water-efficient irrigation methods (e.g. drip irrigation), and to apply practices that improve crop water use efficiency; such as the use of wind breaks, mulching, shading, improved irrigation scheduling, etc.

Linked to resource efficiency is waste minimisation. Food production enterprises should strive to minimise food losses and waste. Studies (e.g. Lundqvist et al. 2008; Gustavsson et al. 2011; Nahman et al. 2012; Institution of Mechanical Engineers 2013) have shown that between 30% and 50% of all food produced for human consumption is lost or wasted along the food supply chain (from production on farm, to consumption at the household level). Not only does this have a negative impact on food security, but it implies that huge quantities of resources embedded in food production are wasted. Reducing food losses and waste is therefore imperative in order to address both food security and resource efficiency. In developing regions, such as sub-Saharan Africa, the majority of food losses occur in the pre-consumer stages of the supply chain, particularly during agricultural production, post-harvest handling and storage, and processing and packaging (Gustavsson et al. 2011; Oelofse and Nahman 2013; Nahman and De Lange 2013). Importantly, vast amounts of energy, water and other resources are utilised in production of food that is ultimately lost or wasted, implying that these resources literally go to waste (FAO 2013; Oelofse 2014). Reducing food losses and waste would therefore go a long way to addressing food security, whilst drastically reducing the use of resources.

In addition, agricultural enterprises should strive to reuse or recycle resources as much as possible, e.g. by using organic waste to make compost, which can then be used to fertilise soils, while improving soil moisture retention; and thereby reducing the need for both water and fertiliser inputs.

3.3.2 "Low Carbon" Agriculture

While the term 'low carbon' economy is generally understood as referring to an economy that reduces emissions of carbon dioxide (CO_2), the underlying principle should clearly be seen as equally relevant to other greenhouse gases, such as methane (CH_4) and nitrous oxide (N_2O), rather than only CO_2 per se. While estimates vary, agriculture as a sector is responsible for approximately 13% of global greenhouse gas emissions (on-farm food production only) (World Resources Institute WRI 2014); rising to as high as 29% if the entire food system is taken into account (CGIAR 2014). The agriculture sector is therefore the second largest contributor to global greenhouse gas emissions, after energy (WRI 2014); and the largest contributor of non-CO_2 greenhouse gases (CGIAR 2014). This excludes emissions from land use change, such as the clearing of forests for expansion of agricultural land (WRI 2014). As such, agriculture will have to reduce its greenhouse gas emissions substantially and contribute to climate change mitigation if it is to qualify as a green economic activity.

In the first place, it is important to identify all the potential sources of greenhouse gas emissions within an agricultural enterprise. Generally speaking, the bulk of the emissions from the agriculture sector are associated with CH_4 from livestock (not relevant to crop production), and N_2O from soils (including natural processes and the application of fertilisers) (WRI 2014). Furthermore, an agricultural enterprise could conduct a more detailed and specific 'carbon footprinting' exercise to identify additional sources of greenhouse gas emissions. These include both direct (scope 1) emissions associated with, for example, the use of fuel for machinery such as tractors, and emissions resulting from the use of chemical fertilisers and pesticides; as well as indirect emissions associated with the use of electricity (scope 2); and emissions arising further upstream or downstream in the supply chain; such as emissions embodied in the production of inputs, or in downstream processing, for example (scope 3).

Thereafter, practices and activities which reduce greenhouse gas emissions can be identified. Some changes in agricultural practices can result in a reduction in greenhouse gas emissions; for example improvements in soil and nutrient management, reducing tillage, improved energy efficiency, and the use of alternative energy sources such as biomass (e.g. biogas derived from anaerobic digestion of crop residues), solar or wind. In addition to mitigation, it will also be important for agriculture to be able to adapt to a changing climate, and to become resilient to climate-related risks such as droughts, floods, and heat-waves. For example, in cases where climate models suggest a decrease in rainfall patterns, it may be necessary to switch toward planting more drought-tolerant crops.

3.3.3 Agriculture Supporting Environmental Protection

Agricultural projects should take special precautions to minimise environmental risks that are associated with the practice of agriculture; such as pollution from fertilisers and pesticides, the degradation of soils through erosion, and the depletion of soil nutrients and soil carbon. In addition, agriculture should aim to support the protection of biodiversity and ecosystems. According to the FAO (2018a, b), a number of agricultural practices can be applied in order to protect biodiversity. Chief amongst these is to ensure a high diversity of crops (avoiding monocultures/mono-cropping); and in particular to ensure the inclusion of perennial crops, which can provide essential habitat for pollination species and natural predators of pests (thereby ensuring a natural cycle of biological control against insects and weeds), and which require less fertiliser application, thereby reducing runoff and pollution, and preventing algal blooms, which are detrimental to aquatic biodiversity. It is also important to maintain a high level of crop-genetic diversity, both on the farm as well as in seed banks. This will contribute towards increasing and sustaining production levels and nutritional diversity through a broad range of agro-ecological conditions, which will increase resilience to changing conditions (FAO 2018a).

The following sections provide further details regarding climate change challenges in the context of agriculture (including the impacts of climate change on water resources), the environmental impacts of agriculture, and recommendations regarding the selection of methods and practices in order to balance the requirements of the agricultural sector with those of the green economy.

3.4 Climate Change Challenges for the Agriculture Sector—Implications for Green Economy Implementation

By now the world is no stranger to the challenges that climate change will bring; including shifting climatic patterns; increased frequency and magnitude of extreme weather events such as droughts, floods, and heat waves. There are also issues relating to human and institutional capacity to adapt to and respond to the effects of climate. Climate change has significant implications for the agriculture sector as climate is a key factor in agriculture. The agriculture sector is facing challenges which include the need to increase the production of food and other commodities for an increasing population (Tubiello et al. 2008). Furthermore, the amount of usable land for agriculture is declining, while the number of people who need to reap benefits from the land is increasing (Brown et al. 2017). The challenges in the agriculture sector are being exacerbated by climate change induced increases in temperatures, rainfall variation and the frequency and intensity of extreme weather events (OECD 2015). Climate change is expected to reduce the productivity of both crop and livestock production systems, due to changes in temperatures, crop water requirements and water availability and quality (OECD 2015; UNEP 2012). The Intergovernmental Panel on Climate Change (IPCC) (2014a, b) notes that climate-related impacts are already reducing crop yields in some parts of the world, a trend that is projected to continue as temperatures rise further.

Changing climatic conditions affect crops and the natural resources that are necessary for their production in various ways. Climate change impacts on crop production are highlighted in Table 3.1; and include reduction in crop yields due to lowering of soil moisture and increased pest incidences due to increasing temperature (Carter and Gulati 2014).

Climate change is also impacting negatively on the resource base on which agriculture depends as it is contributing to resource problems such as water scarcity, pollution and soil degradation (OECD 2015; WFP 2018). UNEP (2012) notes a likely increase in the occurrence of droughts, flooding and increased water scarcity in Africa due to climate change. While the changes might be detrimental for water

Table 3.1 Impact of climate change on crop production (adapted from Carter and Gulati 2014)

Impact of climate change	Direct consequences	Indirect consequences
Average temperature increase	• Reduced quantity and reliability of yields • Increased susceptibility to crop burning • Increased evapotranspiration • Destruction of crops due to increased prevalence and incidence of pests	• Intensified competition for water between sectors due to increased evaporation and decreased water balance • Increased evapotranspiration resulting in reduced soil moisture and reduced crop productivity
Change in rainfall amount and patterns (frequency and intensity)	• Reduced crop quantity and quality • Reduced water availability for crops due to decrease in water resources, decrease in run-off/stream flow • Increased reliance on irrigation • Increased energy consumption for irrigation and crop-spraying systems	• Reduced crop productivity due to soil erosion • Increased probability of fire • Poor quality of crops due to decline in water quality and quantity
Increased severity of drought	• Decreased crop yields • Trade-offs for crop production as water reservoirs come under pressure to meet residential and commercial needs	• Reduced crop productivity due to moisture stress
Increased frequency and intensity of heavy rainfall events	• Increased land degradation and desertification • Damage to crops and food stores • Soil erosion • Water logging—inability to cultivate land • Damage to infrastructure	• Reduced crop productivity due to increased soil erosion

availability in some areas in Africa, in other areas they may alleviate water stress (Schulze 2012). It is imperative that green economy projects are cognisant of the precarious water situation in many areas, and incorporate strategies to conserve and use water efficiently.

Climate change is already threatening the ability of some rain-fed agriculture-dependent regions to maintain levels of agricultural production and food security; and is destabilizing markets (WFP 2018). Extreme weather events are affecting agriculture negatively, for example droughts experienced in southern Africa—see Box 3.1.

The developing world, and in particular Africa, has been identified as being especially vulnerable to the effects of climate change. Weber et al. (2018) note that Africa is supposed to be a climate change hot spot with a high exposure to future climate changes and a low adaptation capacity resulting in a very large vulnerability to future

The October 2015 to March 2016 rainfall season saw the worst drought in southern Africa in the last 35 years. According to the Regional Interagency Standing Committee for southern Africa (RIASCO), the El Niño induced drought negatively affected rain-fed agriculture and caused widespread and severe water and food shortages. The drought caused a second consecutive failed harvest, with a regional maize production shortfall of 9.3 million tons, (RIASCO 2017). Rain-fed agriculture accounts for the livelihoods of many southern Africans, while about 95% of farmed land in sub-Saharan Africa is under rain-fed agriculture (IMWI 2018). RIASCO reports that the situation became so severe that governments could no longer cope individually, and as a result, international assistance was called for culminating in the RIASCO Action Plan 2016/2017(RIASCO 2017). The Action Plan specifically supported five countries who declared national emergencies: Lesotho, Zimbabwe, Swaziland and Malawi. Mozambique declared a Red Alert (highest level of emergency) (RIASCO 2017).

Source: RIASCO 2017

Box 3.1 Case example: El Niño—Induced drought in southern Africa 2015–2017

climate change. Some of Africa's climate vulnerabilities include food and water security risks and degradation of natural resources including irreversible biodiversity loss (IPCC 2001). The continent already experiences a major deficit in food production in many areas, and potential declines in soil moisture will be an added burden (IPCC 2001). Africa's climate change vulnerability is exacerbated by its rapid population growth. According to the UN (2015), 1.3 billion people are projected to be added to the population of Africa by 2050. This increase in population will also have significant consequences for the demand for food and other agricultural commodities and for Africa's ability to meet this demand.

Given the range and magnitude of the development constraints and challenges facing most African nations, the overall capacity for Africa to adapt to climate change is low (IPCC 2001). In addition, factors such as soil degradation, a relatively high dependence on rain-fed agriculture and on natural resources and ecosystems; high poverty rates; limited access to human capital; low levels of preparedness to the effects of climate change; and old and failing infrastructure, especially in rural areas aggravate Africa's vulnerability to climate change (Sibanda et al. 2017). Furthermore, Africa's diversity in climate, landform, biota, culture and economic circumstances, makes it difficult to predict the effects of climate change and the nature and level of adaptation response, thus making Africa particularly vulnerable to climate change impacts (Niang et al. 2014). IFAD (2011) also reports that African farmers are further constrained by the limited functioning of markets, and by prohibitive trade policies, which restricts their access to inputs and markets.

Although agriculture is negatively affected by climate change, it also contributes to climate change as it is responsible for a significant amount of the greenhouse gas (GHG) emissions that are causing climate change (OECD 2015). Green economy implementation in the agriculture sector has to address both the impacts of climate change on agriculture and the contribution of agriculture to climate change and has to incorporate climate change adaptation and mitigation strategies. Climate smart agriculture (described in Sect. 1.4 of this book) incorporates these strategies. Such strategies have to be targeted at farmers in order to build climate resilience and enhance their capacity to sustain agriculture in the face of climate change. OECD (2015) notes the need for farmer targeted climate change response initiatives which should enhance farmer capacity to achieve sustainable productivity growth through climate change mitigating and adaptive practices.

Other enabling factors for appropriate climate change response by farmers include relevant and up-to-date information on risk management and resource use efficiency; and these can stimulate take-up of innovative technologies that support sustainable and climate-friendly goals (OECD 2015). For risk management, access to tools such as weather forecasting or early warning systems enables farmers to take pre-emptive actions to minimise the negative effects of extreme events. Training and education about changing climate conditions and the long-term viability of different agricultural practices help farmers and other stakeholders to make informed investments in adaptation and mitigation (OECD 2015). Building farmer capacity to mitigate and adapt to changing climatic conditions should be central to green economy implementation.

In addition to project or farm level interventions such as farmer capacity building to address climate change, interventions are also required at government level e.g. infrastructure such as dams to store water and policies and taxes to provide guidance and incentives. OECD (2015) notes the important role of governments in provision of infrastructure and policy level interventions that are required for climate change adaptation. What this implies is that green economy implementation does not take place in isolation—it has to be connected to and supported by governments.

Changes in the mean climate away from current states may require adjustments to current practices in order to maintain productivity, and in some cases the optimum type of farming may change (Gornall et al. 2010). These adjustments may present new opportunities for agricultural development. Tubiello et al. (2008) note that the challenges presented by climate change also offer the potential to develop and promote food and livelihood systems that have greater environmental, economic and social resilience to risk. Green economy implementation needs to be aware of potential opportunities from climate change, and project implementers have to be adaptable and be ready to explore new and different trajectories for agriculture.

There are two areas of particular significance for developing regions like Africa in terms of climate change and agriculture, and therefore for green economy implementation; namely ensuring security of production and water security. Adopting a green economy approach and implementing green economy projects in the agriculture sector is central to strengthening both production and water security in the face of climate change. In addition, adopting a green economy approach would facilitate realization of the climate change mitigation potential of the sector. According to OECD (2011), a business as usual approach, as opposed to a green economy approach, will culminate in a future where food security is put at risk due to natural resource limits being exceeded. In order to ensure that this does not happen, OECD (2011) argues for a green-growth path that seeks to identify and implement good policies and embrace opportunities. Climate change will have far reaching impacts on agriculture in Africa through its impacts on food and water security. It is imperative that agricultural green economy projects address these issues; both in terms of contributing to climate change mitigation, as well as adapting to the impacts of climate change.

3.5 Aligning Agricultural Practices with the Requirements of a Green Economy: Choices of Methods and Practices

There are potential tensions and trade-offs between agriculture as a primary sector, which generally has negative impacts on the environment, and the green economy imperatives related to reducing environmental risks. Despite the limitations of agriculture in terms of its negative environmental impacts and the threats posed by resource scarcity, environmental degradation and changing global conditions; agriculture can contribute to a green economy if it adopts practices that align with green economy principles. Adopting such practices requires balancing agricultural and green economy objectives. Agricultural practices determine how agriculture impacts on the biophysical environment, and they also determine agriculture's socio-economic outcomes. Practices which undermine the integrity of the natural resource base pose risks to the environment and to long term agricultural productivity and sustainability while practices which improve the natural resources on which agriculture depends have positive outcomes for both agriculture and the environment.

Sustainable agriculture, which has been defined and discussed in Sect. 1.4 of this book is one of the approaches that can be used to achieve the balance between agricultural and green economy imperatives. This balance is at the core of making agriculture compatible with the green economy. Agroecology is another of the approaches that can align agriculture with green economy principles. Agroecology refers to the application of ecological principles in the design and management of agricultural land. It is a way of identifying the links and interdependencies of various aspects of agricultural ecosystems so that more sustainable production activities can be identified and implemented (InterDev 2015). Agroecological production can regenerate agroecosystems and reverse the damage caused by extractive agricultural production activities (UNEP 2012) and is relevant to green economy implementation as it provides a way of making agriculture work in harmony with the environment.

Adopting green practices and addressing environmental issues will improve the alignment of agriculture with green economy ideals. The greening of agriculture (which has been discussed in Sect. 1.4 of this book) is one of the necessary conditions for aligning agriculture with green economy. However, agriculture's alignment with a green economy requires more than simply 'green agriculture' as there are key social aspects that also need to be considered. In other words, the 'greening' of agriculture is a necessary (but not sufficient) condition for it to be fully aligned with green economy ideals. The greening of agriculture can be achieved through ensuring that agriculture is based on sustainable or agroecological principles; and there is a variety of agricultural production techniques and practices that can achieve this.

Generally, agricultural practices that aim to achieve the following objectives are well suited to aligning agriculture with the green economy:

- Soil improvement (including restoring and enhancing soil fertility);
- Reducing soil erosion and improving the efficiency of water use;
- Reducing chemical pesticide and herbicide use;
- Reducing food spoilage and loss.

(Source: UNEP 2011).

There are several farming methods or techniques that can be applied to achieve these objectives, and that are therefore suitable for application to green economy projects. These include:

- Conservation Agriculture (CA) which has been discussed in Sect. 1.4 of this book,
- Integrated Pest Management (IPM) can help align agriculture with green economy principles. IPM is defined as the careful consideration of all available pest control techniques and subsequent integration of appropriate measures that discourage the development of pest populations, while keeping pesticides and other interventions to levels that are economically justified and reduce or minimize risks to human health and the environment (FAO 2018b). The FAO notes that IPM emphasises the growth of a healthy crop with the least possible disruption to agro-ecosystems, and encourages natural pest control mechanisms (FAO 2018b).
- Organic farming is a production management system that aims to promote and enhance environmental health. It is based on minimising the use of inorganic inputs (such as chemical fertilisers), and represents a deliberate attempt to make the best use of local resources, using methods that minimise adverse effects on the environment and on people. (http://www.ifoam.org/en/organic-landmarks/principles-organic-agriculture). Organic farming also has clear standards and certification to help ensure that its core practices are adhered to. The objectives of organic farming are aligned with environmental and socio-economic green economy principles.
- Climate smart agriculture (CSA) has been discussed in Sect. 1.4 of this book and is one of the practices that can improve alignment of agriculture with green economy principles.
- Market branding certifications such as Good Agricultural Practices (GAP) and others can also contribute to the greening of agriculture and its alignment with green economy principles.
- Several farming practices and technologies can mitigate climate change by reducing greenhouse gas (e.g. carbon dioxide, methane and nitrous oxide) emissions, enhancing carbon storage in soils and plants, and preserving existing soil carbon and thus align agriculture with some of the environmental green economy principles. The use of biochar is an option that has been advanced for carbon dioxide emission reduction in agriculture through carbon sequestration. Biochar is a fine-grained and porous substance produced by the burning of biomass under limited oxygen conditions (Sohi et al. 2009). When added to the soil, biochar stores carbon

for a significantly longer period than would have occurred if the original biomass had been left to decay (Woolf et al. 2010). Incorporating biochar into the soil has been reported to increase fertility, help crop growth, and improve other soil properties (Glaser et al. 2002; Pandian et al. 2016). The enhanced productivity from increased crop growth is a positive feedback that further enhances the amount of carbon dioxide removed from the atmosphere (Woolf et al. 2010).

3.6   Conclusion

In green economy project implementation, biophysical and environmental factors that have implications for crop production have to be considered and balanced with the principles of a green economy. This chapter explores the relevant biophysical and environmental considerations in the context of crop production, and their implications for the ability of crop production initiatives to meet green economic objectives. In particular, the principles of a green economy emphasise that projects must be 'low carbon', use resources efficiently and support environmental protection. The environmental impacts of crop production and their implications for a green economy are also addressed in this chapter. The chapter also discusses climate challenges for agriculture in Africa and South Africa, and implications for a green economy and its implementation. The alignment of agricultural practices with the environmental objectives of a green economy including making adjustments to and selecting appropriate practices and methods are covered.

Chapters one to three covered general issues relating to agriculture and the green economy. In addition, the chapters included direct references to examples from the developing world, Africa and South Africa. This has been done to provide a context for chapters four and five which address practical project level considerations in green economy implementation in the agriculture sector.

References

Allen C (2012) A Guidebook to the Green Economy. Issue 2: Exploring Green Economy Principles. United Nations Department of Economic and Social Affairs (UNDESA): United Nations Division for Sustainable Development

Attwood S, Estrada-Carmona N, Gauchan D, DeClerck F, Wood S Bai K, van ZonneveldM (2017) Using agricultural biodiversity to provide multiple benefits in sustainable farming systems. Biodiversity International. https://www.bioversityinternational.org/fileadmin/user_upload/online_library/Mainstreaming_Agrobiodiversity/Summary_Mainstreaming_Agrobiodiversity.pdf. Accessed 05 Oct 2018

Brown B, Nuberg I, Llewellyn R (2017) Negative evaluation of conservation agriculture: perspectives from African smallholder farmers. Int J Agric Sustain 15:467–481

Busari MA, Kukal SS, Kaur A, Bhatt R, Dulazi AA (2015) Conservation tillage impacts on soil, crop and the environment. Int Soil Water Conserv Res 3:119–129

Carter S, Gulati M (2014) Climate change, the food energy water nexus and food security in South Africa. Understanding the food energy water nexus. WWF-SA, South Africa

Cassman KG, Dobermann A, Walters DT, Yang H (2003) Meeting cereal demand while protecting natural resources and improving environmental quality. Annu Rev Environ Resour 28:315–358

CGIAR (2014) Big facts: focus on food emissions. https://ccafs.cgiar.org/blog/big-facts-focus-food-emissions#.W3aTasL-vnh. Accessed 17 Aug 2018

Crauford PQ, Wheeler TR (2009) Climate change and the flowering time of annual crops. J Exp Bot 60:2529–2539

Delgado A, Gómez JA (2016) The soil. Physical, chemical and biological properties. In: Villalobos FJ, Fereres E (eds), Principles of Agronomy for Sustainable Agriculture. https://doi.org/10.1007/978-3-319-46116-8_2

FAO (1987) Soil quality considerations in the selection of sites for aquaculture. http://www.fao.org/docrep/field/003/AC172E/AC172E00.htm#TOC. Accessed 5 Oct 2018

FAO (2004) Building on gender, agrobiodiversity and local knowledge. http://www.fao.org/3/a-y5609e.pdf. Accessed 30 Oct 2018

FAO (2013) Food Wastage Footprint: Impacts on natural resources. Technical Report. Food and Agriculture Organisation of the United Nations, Rome

FAO (2018a) How to manage biodiversity for food and agriculture. http://www.fao.org/agriculture/crops/thematic-sitemap/theme/spi/scpi-home/managing-ecosystems/biodiversity-and-ecosystem-services/bio-how/en/. Accessed 17 Aug 2018

FAO (2018b) AGP-Integrated Pest Management. http://www.fao.org/agriculture/crops/thematic-sitemap/theme/pests/ipm/en/. Accessed 27 Sept 2018

Glaser B, Lehmann J, Zech W (2002) Ameliorating physical and chemical properties of highly weathered soils in the tropics with charcoal-a review. Biol Fertil Soils 35:219–230

Gomiero T (2016) Soil degradation, land scarcity and food security: reviewing a complex challenge. Sustainability 8:281. https://doi.org/10.3390/su8030281

Gornall J, Betts R, Burke E, Clark R, Camp J, Willett K, Wiltshire A (2010) Implications of climate change for agricultural productivity in the early twenty-first century. Philos Trans R Soc B: Biol Sci 365:2973–2989

Gupta NK, Gupta S, Kumar A (2001) Effect of water stress on physiological attributes and their relationship with growth and yield of wheat cultivars at different stages. J Agron Crop Sci 186:55–62

Gurian-Sherman D (2012) High and dry: why genetic engineering is not solving agriculture's drought problem in a thirsty world. UCS Publications, Cambridge

Gustavsson J, Cederberg C, Sonesson U, van Otterdijk R, Maybe A (2011) Global food losses and food waste: extent, causes and prevention. Study conducted for the International congress SAVE FOOD! At Interpack 2011, Düsseldorf, Germany. Food and Agriculture Organization of the United Nations, Rome

Hatfield JL, Sauer TJ, Cruse RM (2017) Soil: The Forgotten Piece of the Water, Food, Energy Nexus. Adv Agron 143:1–46

Hochman Z, Carberry PS, Robertson MJ, Gaydond DS, Bell LW, McIntosh PC (2013) Prospects for ecological intensification of Australian agriculture. Eur J Agron 44:109–123

Hosonuma N, Herold M, De Sy V, De Fries RS, Brockhaus M, Verchot L, Angelsen A, Romijn E (2012) An assessment of deforestation and forest degradation drivers in developing countries. Environ Res Lett 7:044009

IAASTD (2009) Agriculture at a crossroads: sub-Saharan Africa (SSA) report (vol V). International assessment of agricultural knowledge, science and technology for development. Island Press, Washington, DC

IFAD (2011) Rural poverty report: New realities, new challenges https://reliefweb.int/report/world/rural-poverty-report-2011-new-realities-new-challenges-new-opportunities-tomorrows. Accessed 20 Aug 2018

IFAD (2013) Smallholders, food security and the environment. http://allafrica.com/download/resource/main/main/idatcs/00061832:51f91900626fd396e5e98801329c2358.pdf. Accessed 28 June 2013

IMWI (2018) Rainfed agriculture – summary. http://www.iwmi.cgiar.org/issues/rainfed-agriculture/summary/. Accessed 20 Aug 2018

Institution of Mechanical Engineers (2013) Global food: waste not want not. https://www.imeche.org/docs/default-source/default-document-library/global-food—waste-not-want-not.pdf?sfvrsn=0. Accessed 17 Aug 2018

InterDev (2015) Agroecological farming systems. http://www.fao.org/docs/eims/upload/207703/fiche%20WP3.pdf. Accessed 24 Sept 2018

IPCC (2001) Climate change 2001: impacts, adaptation, and vulnerability. Contribution of working group II to the third assessment report of the intergovernmental panel on climate change. Cambridge University Press. http://www.ipcc.ch/ipccreports/tar/wg2/index.php?idp=378. Accessed 8 Oct 2018

IPCC (2014a) AR5: key findings on implications for agriculture. https://unfccc.int/news/latest-ipcc-science-on-implications-for-agriculture. Accessed 8 Oct 2018

IPCC (2014b) Climate change 2014: synthesis report. In: Pachauri RK, Meyer LA (eds) Contribution of working groups I, II and III to the fifth assessment report of the intergovernmental panel on climate change (Core Writing Team). IPCC, Geneva, Switzerland. http://www.ipcc.ch/pdf/assessment-report/ar5/syr/AR5_SYR_FINAL_All_Topics.pdf. Accessed 20 Aug 2018

Janzen HH, Fixen PA, Franzluebbers AJ, Hattey J, Izaurralde RC, Ketterings QM, Lobb DA, Schlesinger WH (2011) Global prospects rooted in soil science. Soil Sci Soc Am J 75:1–8

Jones RJA, Thomasson AJ (1993) Effects of soil-climate-system interactions on the sustainability of land use: a European perspective. In: Utilization of soil survey information for sustainable land use. Proceedings of the eighth international soil management workshop, pp 39–52. USDA Soil Conservation Service National Soil Survey

Karlen DL, Cambardella CA, Kovar JL, Covin TS (2013) Soil quality response to long-term tillage and crop rotation practices. Soil Tillage Res 133:54–64

Karlen DL, Rice CW (2015) Soil degradation: will humankind ever learn? Sustainability 7:12490–12501

Keane T (2001) Meteorological data—types and sources. In: Holden NM (ed) Agro-meteorological modelling—principles, data and applications. Agmet, Dublin, Ireleand

Köppen W (2011) The thermal zones of the Earth according to the duration of hot, moderate and cold periods and to the impact of heat on the organic world. Meteorol Z 20:351–360

Laidlaw (2009) The effect of soil moisture content on leaf extension rate and yield of perennial ryegrass. Ir J Agric Food Res 48:1–20

Lal R, Stewart BA (1995) Managing soils for enhancing and sustaining agricultural production. In: Lal R, Stewart BA (eds) Soil management: experimental basis for sustainability and environmental quality. CRC Press, Florida

Lanz B, Dietz S, Swanson T (2018) The Expansion of Modern Agriculture and Global Biodiversity Decline: An Integrated Assessment. Ecological Economics 144:260–277

Lundqvist J, de Fraiture C, Molden D (2008) Saving water: from field to fork – curbing losses and wastage in the food chain. SIWI policy brief. Stockholm International Water Institute (SIWI), Stockholm

Luo Q (2011) Temperature thresholds and crop production: a review. Clim Change 109:583–598

Mancosu N, Snyder RL, Kyriakakis G, Spano D (2015) Water scarcity and future challenges for food production. Water 7:975–992

Marvier M (2001) Ecology of Transgenic Crops. Am Sci 89 (2):160

McCauley A, Jones C, Jacobsen J (2005) Basic soil properties. http://landresources.montana.edu/swm/documents/Final_proof_SW1.pdf. Accessed 5 Oct 2018

MEA (2006) Ecosystems and human well-being: Synthesis. World Resources Institute, Washington

Meadows ME, Hoffman MT (2002) The nature, extent and causes of land degradation in South Africa: legacy of the past, lessons for the future. Area 34:428–437

Morison JI, Baker N, Mullineaux P, Davies W (2008) Improving water use in crop production. Philos Trans R Soc B: Biol Sci 363:639–658

Nahman A, De Lange W (2013) Costs of food waste along the value chain: evidence from South Africa. Waste Manag 33:2493–2500

Nahman A, De Lange W, Oelofse S, Godfrey L (2012) The costs of household food waste in South Africa. Waste Manag 32:2147–2153

Natural Capital Forum (2018) What is natural capital? https://naturalcapitalforum.com/about/. Accessed 28 Sept 2018

NEPAD (New Partnership for Africa's Development) (2002) Comprehensive Africa Agriculture Development Programme. FAO, Rome

Niang IOC, Ruppel MA, Abdrabo A, Essel C, Lennard J, Padgham, Urquhart P (2014) Africa. In: Barros VRCB, Field DJ, Dokken MD, Mastrandrea KJ, Mach TE, Bilir M, Chatterjee KL, Ebi YO, Estrada RC, Genova B, Girma ES, Kissel AN, Levy S, MacCracken PR, Mastrandrea, White LL (eds) Climate change 2014: impacts, adaptation, and vulnerability. Part B: regional aspects. Contribution of working group II to the fifth assessment report of the intergovernmental panel on climate change. Cambridge University Press, Cambridge, United Kingdom and New York, NY, USA, pp 1199–1265

OECD (2011) A green growth strategy for food and agriculture www.oecd.org/greengrowth/sustainable-agriculture/48224529.pdf. Accessed 20 Aug 2018

OECD (2015) Agriculture and Climate Change. OECD Trade and Agriculture Directorate https://www.oecd.org/tad/sustainable-agriculture/agriculture-climate-change-September-2015.pdf. Accessed 8 Oct 2018

Oelofse S (2014) Food waste in South Africa: understanding the magnitude: water footprint and cost. In: The vision zero waste handbook. https://issuu.com/alive2green/docs/waste_v4_web. Accessed 17 Aug 2018

Oelofse S, Nahman A (2013) Estimating the magnitude of food waste generation in South Africa. Waste Manag Res 31:80–86

Ontel I, Vladut A (2015) Impact of drought on the productivity of agricultural crops within the Oltenia Plain, Romania. Geogr Pannonica 19:9–19

Pandian K, Subramaniayan P, Gnasekaran P, Chitraputhirapillai S (2016) Effect of biochar amendment on soil physical, chemical and biological properties and groundnut yield in rainfed Alfisol of semi-arid tropics. Arch Agron Soil Sci 62:1293–1310

Porter J, Gawith M (1999) Temperatures and the growth and development of wheat: a review. Eur J Agron 10:23–36

RIASCO (2017) UN office for the coordination of humanitarian affairs. http://reliefweb.int.report/world-riasco-action-plan. Accessed 20 Aug 2018

Schulze RE (2012) A 2011 perspective on climate change and the South African water sector. Water research commision report: WRC report No. TT 518/12

Scotcher JSB (2009) The green choice living farms reference 2009/2010 version. In: Goldblatt, A (ed) Unpublished report to Green Choice (a WWF and Conservation International partnership)

Sibanda LM, Mwamakamba SN, Mentz M, Mthunzi T (eds) (2017) Policies and practices for climate-smart agriculture in sub-Saharan Africa: a comparative assessment of challenges and opportunities across 15 countries. Food, Agriculture and Natural Resource Policy Analysis Network (FANRPAN), Pretoria

Sohi S, Lopez-Capel E, Krull E, Boll R (2009) Biochar, climate change and soil: A review to guide future research. CSIRO Land and Water Science Report series, ISSN: 1834-6618

Tilman D, Cassman KG, Matson PA, Naylor R, Polasky S (2002) Agricultural sustainability and intensive production practices. Nature 418:671–677

Tubiello F, Schmidhuber J, Howden M, Neofotis PG, Park S, Fernandes E, Thapa D (2008) Climate change response strategies for agriculture: challenges and opportunities for the 21st Century. The World Bank. http://siteresources.worldbank.org/INTARD/Resources/dp42Combined_web.pdf. Accessed 8 Oct 2018

Tubiello FM, Salvatore M, Ferrara AF et al (2015) The contribution of agriculture forestry and other land use activities to global warming, 1990–2012. Glob Change Biol 21:2655–2660

Tutwiler A, Bailey A, Attwood S, Remans R (2017) Why mainstream agricultural biodiversity in sustainable food systems? Biodivers Int. https://www.bioversityinternational.org/fileadmin/

user_upload/online_library/Mainstreaming_Agrobiodiversity/Summary_Mainstreaming_ Agrobiodiversity.pdf. Accessed 5 Oct 2018

UN (2015) United nations, department of economic and social Affairs, population division world population prospects: The 2015 revision, key findings and advance tables. Working paper no. ESA/P/WP.241

UNEP (2011) Towards a Green Economy: Pathways to sustainable development and poverty eradication. United nations environment programme, Nairobi. https://www.cbd.int/financial/doc/ green_economyreport2011.pdf. Accessed 16 Oct 2018

UNEP (2012) Principles for a green, fair and inclusive economy Version 3. http://www.unep.org/ greeneconomy/Portals/88/documents/GEI%20Highlights/Principles%20of%20a%20green% 20economy.pdf. Accessed 15 Sept 2013

Vermeulen SJ, Campbell BM, Ingram JSI (2012) Climate change and food systems. Annu Rev Environ Resour 37:195–222

Weber T, Haensler A, Rechid D, Pfeifer S, Eggert B, Jacob D (2018) Analyzing regional climate change in Africa in a 1.5, 2, and 3°C global warming world. Earth's Future 6:643–655

WFP (2018) Climate impacts on food security. https://www.wfp.org/climate-change/climate-impacts. Accessed 20 Aug 2018

Woolf D, Amonette JE, Street-Perrott FA, Lehmann J and Joseph S (2010) Sustainable biochar to mitigate global climate change. Nat Commun 1:56. https://doi.org/10.1038/ncomms1053

WRI (World Resources Institute) (2014) Everything you need to know about agricultural emissions. http://www.wri.org/blog/2014/05/everything-you-need-know-about-agricultural-emissions. Accessed 17 Aug 2018

WWF (2017) Time is ripe for change: towards a common agricultural policy that works for people and nature. WWF Position paper. http://d2ouvy59p0dg6k.cloudfront.net/downloads/wwf_ position_paper_on_cap_post_2020___final__contact_.pdf. Accessed 28 Sept 2018

Chapter 4
Making Sense of Green Economy Imperatives at a Practical Level: Case Studies of Small-Scale Vegetable Production in South Africa

4.1 Theoretical Framework for the Case Studies

There is documented information on crop-based agricultural initiatives in South Africa and elsewhere, for example that available in various agriculture manuals. However, this information does not incorporate green economy considerations, and is therefore not sufficient to fully inform green economy implementation. Field case studies of vegetable production enterprises are used in this study to analyse the practical implementation environment for an agricultural green economy, and to generate information of relevance to the design and implementation of agricultural green economy initiatives. In discussing the field case studies, reference is made to green economy 'projects' in some instances. The term 'project' in this context is used to describe activities focused on production of a specific crop; for example, producing a tomato crop would be referred to as a tomato production project.

As stated in Chap. 1, the international community has identified generic green economy principles (summarised by Allen 2012) which in essence define a green economy. However, the green economy cannot be implemented using a 'one-size-fits-all' approach, but should rather be tailored to fit specific contexts. To incorporate contextual issues, the case studies presented in this chapter were also informed by South Africa's green economy and sustainable development priorities, as well as by local-level issues that would be of relevance for planning, setting up and operating an agricultural green economy project.

In this regard, a key document articulating South Africa's green economy and sustainable development aspirations is the National Development Plan (NDP) (NPC 2011). The NDP (Chap. 5) articulates a vision for South Africa's green economy, stating that "by 2030, South Africa's transition to an environmentally sustainable, climate change resilient, low-carbon economy and just society will be well under way" (NPC 2011: 179). According to the NDP, the transition to a green economy in

© The Author(s), under exclusive license to Springer Nature Switzerland AG 2018   61
C. Musvoto et al., *Green Economy Implementation in the Agriculture Sector*,
SpringerBriefs in Agriculture, https://doi.org/10.1007/978-3-030-01809-2_4

South Africa is to be guided by principles which include justice, ethics, sustainability, protection of ecosystems, full cost accounting and effective participation of social partners (NPC 2011). The NDP emphasises that there is a need to find ways to break the links between economic activity, environmental degradation and carbon-intensive energy consumption.

The case studies were also informed by South Africa's National Strategy for Sustainable Development and Action Plan (NSSD) (DEA 2011). The fundamental principles of the NSSD are human dignity and social equity; justice and fairness; democratic governance; and a healthy and safe environment. South Africa's Green Economy Accord (EDD 2011) also informed the case studies. The focus of the Accord is addressing unemployment, poverty and inequality, with job creation being one of its targets. Support for small-scale agriculture is one of the agriculture-related goals articulated in the Accord; while its commitments include protecting the environment, waste recycling, sharing costs of development fairly across society, promotion of youth employment and skills development; promoting decent work, and supporting the broader goals of a green economy.

Relevance to the local context is a core facet of the green economy. In the context of South Africa, where the green economy is central to social improvement, green economy projects have to address issues that are relevant to communities in the local area, so that the communities in question can benefit. Jackson and Victor (2013) note that a vision for enterprises on which the green economy can be built should be defined in terms of providing the capabilities for people to prosper and for communities to thrive. Some of the operational principles that enterprises should fulfil in the green economy include reflecting the interdependence of social, economic and environmental systems, and having objectives which address these three factors; and encompassing matters of social inclusion and justice, policy, power and governance (Jackson and Victor 2013). Other factors which are critical to a green economy at implementation level as defined by Jackson and Victor (2013) include building capabilities for communities to thrive; maintaining and enhancing social and environmental well-being; and addressing well-being beyond money and material abundance. These factors are relevant at community level in South Africa and were incorporated into the case studies as far as possible.

Taking the above into consideration, a theoretical framework was developed in order to systematically organise, compare and discuss the case studies. The framework integrates the different factors that should be considered in green economy implementation into four socio-economic and two environmental themes (Fig. 4.1) which summarise green economy imperatives in the context of the case studies.

The themes capture green economy principles and other factors of relevance to green economy implementation in the context of small scale vegetable farming in South Africa and are further highlighted in Table 4.1. The themes are as follows:

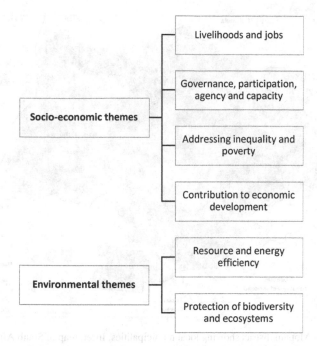

Fig. 4.1 Theoretical framework developed to systematically organise, compare and discuss the case studies

4.1.1 Socioeconomic Themes

Livelihoods and jobs: One of the benefits that a shift in agriculture towards the green economy could bring is more sustainable livelihoods and jobs. The UNDP (2012) notes that employment that is secure and decent is central to reducing poverty, while also supporting more inclusive, equitable and sustainable development. 'Green jobs' are also an important feature of the green economy. In the South African context, green jobs are defined as work that is geared towards contributing to preserving or restoring environmental quality and reducing energy, water and materials consumption (DEA 2007). At the same time, employment opportunities created through the green economy need to adhere to the principles of 'decent work' i.e., they should offer adequate wages, safe working conditions, job security, reasonable career prospects and worker rights (DEA 2007).

Governance, participation, agency and capacity: Good governance, participation, agency[1] and capacity building are key factors in green economy implementation. Farming First (2018) posits that agriculture in the green economy should adopt

---

[1] Agency in this context refers to the power people have to think for themselves and act in ways that shape their experiences and life trajectories. It is also about the ability of an individual to speak freely, raise their opinions and stand up for themselves without fear of retribution or unequal treatment. All of this is also related to empowerment.

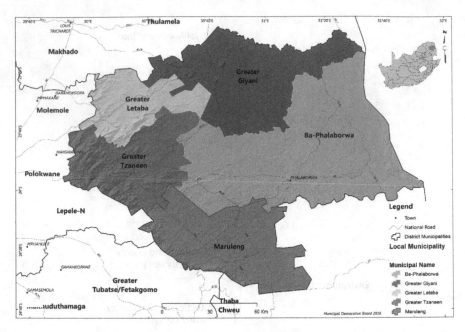

Fig. 4.2 Map of Mopani district showing local municipalities. Inset: map of South Africa showing location of Mopani District (shaded in green)

a knowledge-centred approach, focused on (i) support for knowledge sharing, and advisory and training services; and (ii) productivity supported through innovation and best practices. Drawing on the knowledge and capacity of all involved in the green economy is crucial for success. The UNDP (2008) recognises three levels of capacity, consisting of (i) the individual, (ii) the organisation, and (iii) the enabling environment, which includes the political, social, economic, policy, legal and regulatory systems within which organisations and individuals operate. The UNDP also notes that capacity development is not a 'once-off' intervention, but a process (Table 4.1).

Inequality and poverty: Addressing inequality and poverty is one of the priorities for a green economy, particularly in developing countries. In addition to poverty reduction, agriculture in a green economy context can be an ideal platform for addressing other inequalities, for example those relating to persons with disabilities and gender. The International Policy Centre for Inclusive Growth (Randriamaro 2012) argues that in the context of a changing climate, women's issues in particular should be addressed by the green economy. For example, the general lack of access to and control over land and other natural resources, technologies and credit for women across the globe should be addressed. According to Randriamaro (2012), social inclusivity is another factor that should be addressed through the green economy, so as "to ensure that women and other disadvantaged groups are given specific

Table 4.1 Themes used to organise the case studies and the factors which informed each theme

Theme	Informing factors					
	Green economy principles	National development plan	National strategy for sustainable development	Green economy accord	Relevance to the local context (Source: Jackson and Victor 2013)	
Socio economic themes	Livelihoods and jobs	Decent work and green jobs	Unemployment	Human dignity	Youth employment; Decent work	People prosperity; Thriving communities; Well-being beyond money and material abundance
	Governance, participation, agency and capacity	Governance and rule of law; Democratic; Participatory	Ethics; Effective participation	Democratic governance	Skills development	Capability for people to prosper; Power; Governance; Social well-being
	Inequality and poverty	Equity, fairness and justice	Justice; Poverty	Social equity; Justice; Fairness	Poverty; Inequality	Social inclusion; Justice
	Contribution to economic development	Sustainable development	Transition to a green economy	Sustainable development	Green economic development	Thriving communities; People's prosperity
Environmental themes	Resource and energy efficiency	Resource and energy efficiency; Low carbon	Protection of ecosystems	Efficient use of natural resources	Waste recycling	Inter-dependence of social, economic and environmental systems
	Protection of biodiversity and ecosystems	Protect biodiversity and ecosystems; Respect planetary boundaries, and ecological limits	Low carbon; Climate resilience; sustainability	Healthy and safe environment; Effective climate change responses	Environmental protection	Environmental well-being

attention, not only as the main victims of the negative impacts of climate change and environmental degradation, but also as central agents in achieving sustainable development". Randriamaro's (2012) comment here specifically speaks to agency within these vulnerable groups. However, without appropriate social policies, the green economy could exacerbate existing gender inequities related to gendered patterns of labour, gender-segregated employment patterns, and discrimination (Randriamaro 2012).

Contribution to economic development: In South Africa, agriculture in a green economy context is expected to be a key driver of rural development (DPME 2014). For this to happen, the social, economic and biophysical environments at local and country level have to be conducive for green economic development. For example, an enabling environment for the green economy in agriculture would include access to markets. Having access to markets is essential for enabling farmers to contribute to the economic development of a region as farmers need to be able to get their produce to the market, and to receive equitable price treatment when they do (Farming First 2018). In addition, their ability to provide work for people, thus supporting livelihoods, is an important part of this contribution. The ability of the agriculture sector to provide green jobs is already proven (UNDP 2012), as well as its ability to provide jobs for those who are most vulnerable to climate change (Randriamaro 2012).

4.1.2 Environmental Themes

Resource and energy efficiency: A key attribute of a green economy is resource and energy efficiency. The term 'efficiency' is used to quantify the relative output obtainable from a given input (or, the level of input used to produce a given output). An efficient method of producing a product is that which uses the least amount of inputs or resources to obtain a given amount of the product; or which produces the maximum output from a given level of inputs. In the case of crop production, efficiency entails making better use of resources such as water, land, and fertilisers. One of the four primary goals for agriculture in a green economy according to the World Farmers Organisation is to 'produce more with less' (World Farmers' Organisation 2012). Production efficiency is thus a prerequisite for agriculture in a green economy. Mateo and Ortiz (2013) describe agricultural production efficiency through the concept of "eco-efficiency". Eco-efficient agriculture increases productivity while decreasing negative impacts on natural resources through approaches that meet the economic, social and environmental needs of the rural poor (CIAT 2009).

Protection of biodiversity and ecosystems: Environmental protection is central to the green economy, and is also a recurring theme in South Africa's development strategies. Protection of biodiversity and ecosystems is especially important in the context of agriculture's negative environmental impacts, as discussed in Sect. 3.4 of this book. Given the central role of biodiversity and ecosystem services in maintaining sustained agricultural productivity in the long term, it is critical that agricultural green economy projects are able to enhance and maintain the underlying ecosystems upon which they depend.

4.2   Approach and Methods for the Case Studies

The case studies incorporated a desktop review and qualitative field studies.

4.2.1   The Desktop Review

A desktop review was used to identify all the factors that are relevant to a crop-production based green economy project in general, and in the South African context specifically. A literature review of the green economy, its principles and role in local economic development was conducted. In addition, the position and role of agriculture in a green economy was reviewed. This information was framed within South Africa's development context, which was informed by current strategies and policies.

4.2.2   The Field Studies

The objective of the field studies was to understand the practical aspects of small-scale vegetable production enterprises in the context of a green economy. The studies were conducted in 2014 within a 50 km radius of the town of Tzaneen (23.8333° S, 30.1667° E) in the Limpopo Province of South Africa, and covered seven small-scale vegetable farms. The focus on small-scale vegetable production was informed by documented information on the key potential role of this sector in contributing to a green economy and addressing poverty both in South Africa and in other developing countries (NPC 2011; UNEP 2014). Case study farms were identified through consultation with the local government agricultural advisory services (the Limpopo Department of Agriculture and Rural Development, LDARD). Information was collected through semi-structured interviews with farmers under the six themes discussed in Sect. 4.1 above. The issues investigated under each theme are shown in Table 4.2.

In addition to theme-specific information, general contextual data on each farm was also gathered on aspects such as cropped areas, crops produced, production methods and practices. Tenure, land ownership and general management, decision making and problems faced were also covered. The interviews were augmented with researcher observations conducted on each farm. During these observations, visible evidence relating to both contextual and theme-specific information, such as crops that were in the field, production practices such as irrigation and mulching, conservation structures, and equipment and water resources, were noted.

Table 4.2 The information gathering framework used in the case studies

	Framework Themes	Information gathered on each farm
Socio-economic	Livelihoods and jobs	Employment creation, opportunities for secure livelihoods
	Governance, participation, agency and capacity	Participation and inclusion of people on the basis of factors such as gender, age, and disability; capacity building, knowledge and information provision; networking
	Inequality and poverty	Promoting equality and addressing inequalities through direct and indirect participation in vegetable production
	Contribution to economic development	Marketing and linkages to both local and national input and output markets
Environmental	Resource and energy efficiency	Use of various resources, including fertilisers, energy, and water; focusing on efficiency of use
	Protection of biodiversity and ecosystems	• Production practices in the context of conservation of soils; pollution and waste management; carbon emissions • Awareness of and practices aimed at protecting biodiversity and ecosystems; • Potential environmental impacts of agricultural practices followed • Investments in building natural capital

4.3   Context: Small-Scale Vegetable Production and the Green Economy in Greater Tzaneen, Limpopo Province, South Africa

Small-scale farming is an important livelihood activity. About 2.5 billion people in the world directly derive their livelihoods from small-scale farming (IFAD 2013). Small-scale farmers produce about 80% of Sub-Saharan Africa's food supply (FAO 2012a). In South Africa, small-scale farming is mostly carried out in communal areas of the former Bantustans or homelands,[2] and provides an important source of livelihoods for family members, and occasional employment to others (Aliber and Hart 2009). Small-scale farmers produce food to meet family needs while also supplying local and regional markets, where large numbers of informal traders make a living (Chikazunga and Paradza 2013). In terms of contribution to the country's food supply, small-scale farming in South Africa is not as important as it is in the rest of Africa, contributing only about 5% of national agricultural output (Aliber and

---

[2]The Bantustans or homelands, established by the Apartheid Government were areas to which the majority of the black population was moved to prevent them from living in the urban areas of South Africa. Source: https://www.sahistory.org.za/article/homelands.

Hart 2009). Nevertheless, small-scale farming is an important livelihood activity and contributes significantly to household food supply.

The government of South Africa views small-scale agriculture as a tool for rural development and poverty reduction. South Africa's NDP identifies smallholder agriculture as a potential driver of rural development and as an activity that could improve the livelihoods of at least 370 000 people (NPC 2011). Furthermore, South Africa's government has committed to increase the number of smallholders producing food for sale from about 200 000 to 500 000 during the period 2014 to 2020 (Aliber and Hall 2012).

In developing countries, agriculture is deemed a crucial sector for the green economy (FAO 2012a). In Africa, the importance of natural resource-based sectors such as agriculture and forestry in supporting green economic growth is underscored by the fact that these sectors continue to provide large numbers of jobs (UNECA 2012). South Africa has adopted the principle of green economic growth, with the green economy being prioritised as one of the key economic drivers in South Africa's Medium Term Strategic Framework (MTSF) 2014–2019 (DPME 2014). Small-scale vegetable production has a key potential role in contributing to a green economy and addressing poverty both in South Africa and in other developing countries (NPC 2011; UNEP 2014; UNEP 2016). The envisaged central role of small-scale agriculture in South Africa's green economy is articulated in the country's Green Economy Accord (EDD 2011), which includes several agriculture-related goals.

Small-scale farming is not only critical for food production; it also involves the management of agroecosystems that provide a range of public goods and services. Eighty percent of the farmland in sub-Saharan Africa and Asia is managed by smallholders (working on up to 10 ha) (FAO 2012b). The goods provided by agroecosystems include food, feed, fibres, agrofuels, medicinal products and ornamentals; while the services include stewardship of ecosystems, landscape amenities and cultural heritage (McIntrye et al. 2009). The role of small-scale farming in environmental management makes it particularly relevant to the green economy.

4.3.1 The Biophysical and Socio-Economic Context

The research was conducted in the Greater Tzaneen Local Municipality (GTLM) in the Mopani District of Limpopo Province (Fig. 4.2). The climate in GTLM ranges from tropical to subtropical; with annual rainfall of 729 to 1500 mm, most of which falls during the period September to March (Tzaneen Information 2013). The rainfall patterns, subtropical climate and fertile soils in GTLM allow the production of a wide range of crops, including a variety of vegetables and fruits. According to the municipality's Integrated Development Plans (IDPs) for 2014/15 and 2018/19 (GTLM 2014, 2018), agriculture is an important sector in GTLM, where both commercial and subsistence agriculture are practised.

According to the IDPs (GLTM 2014; 2018), the three sectors which have greatest potential to drive growth of the local economy are agriculture, tourism and agro-processing.

The population of GTLM is mostly rural (82% of the households), and lives on communal land (IDC and SA LED Network 2011); with many residents dependent on subsistence agriculture (GTLM 2018). Most of the land (82%) is classified as communal land and the tenure status of the majority of the residents is secure (StatsSA 2011). Young people between the ages of 14 and 35 constitute 40% of the total population of the municipality, while 48% of households and 50.3% of all agricultural households in the municipality are female-headed (GTLM 2014). The education levels of residents in GTLM are low. According to StatsSA (2011), 42.2% of the residents have "some primary" education, while 37.3% have "some secondary" education. Only 9.9% of the residents completed secondary education (StatsSA 2011). According to StatsSA (2011), 30.1% of individuals who are designated as the head of an agricultural household have no schooling, and only 13.2% have completed high school education. South Africa's latest census figures (StatsSA 2011) indicate that 41% of people living in the GTLM have no source of income; while the unemployment rate is 36.7%. The GTLM IDP for 2018/19 alludes to the need to create decent work and sustainable livelihoods for residents (GTLM 2018). Of those residents who do earn an income, 45% earn below the national average monthly income (StatsSA 2011).

All villages in the GTLM have access to electricity, and 86.2% of residents use electricity as their main source of lighting (StatsSA 2011; GTLM 2018). Currently there is insufficient electrical capacity to allow for expansion of electricity services; while problems with infrastructure maintenance and reliability of electricity supply are increasing, especially in the rural parts of the municipality (GTLM 2018). The municipality is planning for use of renewable energy such as solar and biogas (GTLM 2018). The GTLM has a network of paved major roads (42% of the road network) and gravel roads. However, there are budgetary issues with the maintenance of gravel roads, and maintenance backlogs occur (GTLM 2018).

4.4 Characteristics of Small-Scale Vegetable Farms

All the assessed farms produce a variety of vegetables, including tomatoes, leafy vegetables (spinach, lettuce, cabbage), root vegetables (carrots, beetroot, onion, sweet potato), a variety of peppers, cucumbers, okra, sweetcorn, squashes and green beans. Each farm grows between three and eight different types of vegetables simultaneously and/or in rotations. The vegetables are marketed to national formal markets such as supermarket chains and the Johannesburg and Pretoria (two large cities located approximately 400 km from Tzaneen) fresh produce markets. Local (within a 40 km radius of the farms) formal markets are also used, including supermarkets and establishments such as hotels; with some farms supplying a local tomato processing factory. All the farms also market their produce informally through farm-gate sales to individual consumers, local informal traders and vendors. The produce that is sold to formal markets has to be packaged appropriately (in boxes, pockets and plastic crates), which farmers have to purchase together with various inputs. Organic farmers use inputs such as compost, chicken and cattle manure, organic planting materials (seed and seedlings), organic fertilisers and organic pesticides. Conventional farmers use regular (non-organic) planting materials, chemical fertilisers and pesticides.

Some of the farms use organic production methods (these are referred to as 'organic' farms in this chapter), while others use industrial or conventional production methods (these are referred to as 'conventional' farms). Some farms are held under private tenure while others are under communal tenure. All the vegetables are irrigated, and farms have a reliable water supply. Water shortages were not an issue at the time of conducting the studies; however the water situation may have changed since then. Each farm hires both permanent and casual workers. The casual workers are hired as required, particularly for manual weed control and harvesting. The profile of each farm is shown in Table 4.3.

At the time of conducting the studies all the organic farmers were members of the Nkomamonta Organic Farmers' Cooperative, which was started in 2005. The members of Nkomamonta farm individually and are located in different areas, but use the cooperative as a platform for sharing information on organic production and for accessing markets. Prior to the studies members of Nkomamonta had worked together to get certification as organic producers and to secure marketing contracts with leading supermarket chains in South Africa.

4.5 Alignment with and Opportunities for Green Economy Implementation

Some of the attributes of the case study farms are aligned with green economy imperatives (as articulated through the six themes of the framework described in Sect. 4.1). These attributes can therefore be considered as strengths from a green economy perspective. These areas of alignment indicate opportunities that can be built upon and developed further in green economy implementation. The characteristics of each farm that are aligned with the green economy imperatives are listed in Table 4.4, and discussed below. Areas of misalignment with the green economy are discussed in Sect. 4.6. The discussion of both is structured according to the themes of the framework.

Table 4.3 Characteristics of case study farms

Farm	Cropped area and production practices	Water source and irrigation method	Markets	Tenure arrangements and livelihood opportunities
A	• 17 Hectares • Organic	• 3 Farm dams • Drip irrigation	• National formal • Local formal • Local informal	• Family farm • Private tenure • 14 Permanent employees
B	• 8 Hectares • Organic	• River, dam and 2 boreholes • Drip irrigation	• National formal • Local informal	• Family farm • Private tenure • 6 Permanent employees
C	• 4 Hectares • Organic	• River and borehole • Drip irrigation	• National formal • Local informal	• Family farm • Private tenure • 11 Permanent employees
D	• 3 Hectares • Organic	• Dam • Drip irrigation	• National formal • Local formal • Local informal	• Family farm • Communal tenure • 6 Permanent workers
E	• 1 Hectare • Conventional	• River • Drip irrigation	• National formal • Local informal	• Family farm • Communal tenure • 10 Permanent staff (female only)
F	• 2.5 Hectares • Conventional	• Two dams • Drip irrigation	• Local formal • Local informal	• Women's cooperative (5 members; 4 members aged 18 to 35) • Communal tenure • All members of cooperative work on the farm
G	• 8 Hectares • Conventional	• Dam	• Local informal • National formal • Local formal	• Family farm • Private tenure • 15 Full time workers

Table 4.4 Profile of case study farms highlighting alignment to green economy imperatives (captured through the study themes)

Farm	Livelihoods, jobs and addressing inequality and poverty	Governance, participation, agency and capacity	Contribution to economic development	Resource and energy efficiency	Protection of biodiversity and ecosystems
A (Organic)	• Permanent and casual workers • Majority female workers • Relatively safe work environment—no agrochemicals	• Family farm • Male and female members participate in decision-making • Actively looks for capacity building opportunities • Head of farm has active role in local organic farmer network • Succession plans in place	• Supplies local and national formal markets • Supplies local informal markets	• Drip irrigation • Manure and compost used • Only organic inputs used	• Uses manure and compost • No agrochemicals used • Manual control of invasive alien vegetation • Only uses organic inputs
B (Organic)	• Permanent and casual workers • Majority female workers • Some employees younger than 35 • Relatively safe work environment—no agrochemicals • Direct sales to local traders	• Family co-operative • All members involved in decision-making • Participation of women and men • On the farm training for younger staff • Provision for information sharing through membership of organic farmers' cooperative	• Supplies local and national formal markets • Supplies local informal markets	• Drip irrigation • Manure and compost used • Only organic inputs used	• Uses manure and compost • No agrochemicals used • Uses certified organic inputs

(continued)

Table 4.4 (continued)

Farm	Livelihoods, jobs and addressing inequality and poverty	Governance, participation, agency and capacity	Contribution to economic development	Resource and energy efficiency	Protection of biodiversity and ecosystems
C (Organic)	• Permanent and casual workers • Majority female workers • Relatively safe work environment—no pesticides • Direct sales to local traders	• Run by father/son team • Succession plan in place—son to eventually take over farm • Provision for information sharing through membership of organic farmers' cooperative	• Supplies local and national formal markets • Supplies local informal markets	• Drip irrigation • Manure and compost used • Only organic inputs used	• Uses manure and compost • No agrochemicals used • Uses certified organic inputs
D (Organic)	• Full-time and casual workers • Female workers • Relatively safe work environment—no pesticides • Direct sales to locals	• Provisions for information sharing through membership of organic farmers' cooperative	• Supplies local and national formal markets • Supplies local informal markets	• Drip irrigation • Water flows to farm through gravity—no pumping • Reduced tillage practised • Only organic inputs used	• Reduced tillage—promotes soil organic matter build up and soil biodiversity • Manure and compost used • No agrochemicals used
E (Conventional)	• Full-time and casual workers • Female employees • Direct sales to consumers in village	• Run by father/son team • Succession plan in place—son to eventually take over farm	• Supplies local and national formal markets • Supplies local informal market • Inputs procured from Tzaneen	• Drip irrigation	• No herbicides used

(continued)

Table 4.4 (continued)

Farm	Livelihoods, jobs and addressing inequality and poverty	Governance, participation, agency and capacity	Contribution to economic development	Resource and energy efficiency	Protection of biodiversity and ecosystems
F (Conventional)	• Provides livelihoods and employment for 4 people (cooperative members; including 3 young women (less than 35 years old) • Casual employees hired as required • Direct sales to locals	• Women's cooperative • Skills development through participation and guidance of government advisory services • Skills transfer through formal mentorship • Cooperative run on basis of a constitution • Decision making by consensus	• Supplies local tomato processing factory • Supplies local informal markets • Inputs procured from Tzaneen	• Drip irrigation	• No herbicides used
G (Conventional)	• Full-time and casual workers • Female employees of all ages • Direct sales to locals	• Run by father/daughter team • Succession plan in place—daughter to eventually take over running of farm • Knowledge and information acquisition through membership of Limpopo Tomato Growers' Association	• Supplies national markets • Supplies local formal market and tomato processing factory • Supplies local informal market • Inputs procured from Tzaneen	• Drip irrigation • Crop residues are fed to cattle and manure from the cattle is used as fertiliser • Manure and compost used	• Use of manure and compost • No herbicides used

4.5.1 Livelihoods and Jobs

Each farm produces food and provides livelihood opportunities for its owners and through employment of both full-time and casual workers who are mainly females. The jobs on the farms involve manual labour related to vegetable production such as planting, weeding and harvesting. The farms also supply food directly to local consumers, who buy produce directly from the farms or from vendors supplied by the farms. The organic farms do not use agrochemicals,[3] which eliminates workers' risk of exposure to toxic chemicals and thus provide a relatively safe work environment.

4.5.2 Governance, Participation, Agency and Capacity

Six of the seven farms are family farms, and some are run as family cooperatives. Decision-making on these farms is inclusive and involves both male and female members. Most farms have some form of constitution (as a family cooperative or group cooperative); incorporating some form of governance structure. However, at the time of research there was no clarity on how such governance structures are implemented. There is some networking among the farmers through the Nkoma-monta Farmers' Cooperative and the Limpopo Tomato Growers Association. These networks are also sources of knowledge and information. Farm F is involved in a mentorship agreement with a more experienced farmer, who provides hands-on technical support to the group. All the farmers receive technical and information services from the government agricultural advisory services; although they pointed out that the service needed improvement. NGOs and universities had in the past (prior to the study) provided training on organic vegetable production and marketing.

4.5.3 Inequality and Poverty

The farms mostly employ women; thus providing them with opportunities to earn livelihoods within their communities. In addition, both women and men are involved in decision making on the farms. The participation of women in different aspects of agricultural production resonates with the principles of a green economy. The farms are located in an area with high levels of poverty and unemployment (GTLM 2014; StatsSA 2011). In South Africa, social norms and persistent stereotypes often shape inequitable access to opportunities, resources and power for women and girls; and this manifests in areas such as employment. For example, the unemployment rate in South Africa is higher among females (29.8%) than males (26%) (StatsSA 2018). Poverty rates are higher among unemployed people in South Africa (World Bank 2018).

---

[3]Chemicals used in agriculture, and this includes synthetic or chemical fertilisers and pesticides such insecticides, herbicides and others.

Provision of employment opportunities contributes to addressing both poverty and inequality. One of the farms (F) is a women's cooperative, in which most members are younger than 35. The involvement of young females is particularly aligned with the green economy objective of addressing inequalities and poverty. In South Africa, the unemployment rate for the youth (defined as those aged 15–35) is higher (38.8%) than that for older people 17.9%), regardless of educational attainment (StatsSA 2018).

4.5.4 Contribution to Economic Development

All the farms supply both local and national markets. The local market is supplied through farm-gate sales to informal bulk traders. The informal traders in turn supply a local vending industry, which contributes to the local economy and provides livelihoods for some people. The farms generally procure inputs and supplies within a 40 km radius, and thus support local businesses. The creation of employment opportunities contributes to the local economy. However, this employment is still quite limited, not only in scope but also in terms of sustainability and security.

4.5.5 Resource and Energy Efficiency

All the farms use drip irrigation. If properly designed, installed, and managed, a drip irrigation system can reduce water losses through evaporation and deep drainage in comparison with flood or overhead sprinkler irrigation. As a result, energy is also saved, through reduced pumping of water. A properly installed drip system can save as much as 50 to 80% of the water normally used in other types of irrigation systems; and minimises fertiliser leaching (Aujla et al. 2005; University of Massachussets 2014). Through the improved water efficiency of drip irrigation systems, it can be assumed (no measurements were taken) that energy is saved (all else being equal), as less water has to be pumped. In addition, drip irrigation can eliminate many diseases that are spread through water contact with the foliage (FAO 2001), which can in turn reduce pesticide use. Although not measured in this study, the use of drip irrigation could also be minimising soil nutrient loss through leaching, thus resulting in savings on fertilisers and improving efficiency of production. The organic farms (except one) use compost produced on farm as well as organic fertilisers. None of the organic farms use agrochemicals. Reduced use of agrochemicals translates into reduced risk of pollution, as well as a reduced energy and carbon footprint, and reduced vulnerability to the volatile input costs of agrochemicals. A large amount of energy is used in the manufacture and transportation of fertiliser, especially fertilisers containing nitrogen (N) (Soil Conservation Council of Canada 2001). In addition, some estimates indicate that pesticide manufacturing represents between 6 and 16%

of the energy used in crop production (Audsley et al. 2009). The resource use on organic farms is therefore more aligned with green economy principles than that on conventional farms.

4.5.6   Protection of Biodiversity and Ecosystems

The use of compost and organic fertilisers on organic farms has positive implications for biodiversity, as compost builds up soil organic matter, which in turn promotes soil biodiversity, structure and health. Compost stimulates soil microbial activity and crop growth (Pascual et al. 1997; Zhen et al. 2014). By increasing the soil's ability to hold water and nutrients, compost also changes the soil structure and improves soil health. In some studies, soil biodiversity (measured as microbial biomass[4]) was found to be significantly increased through the application of some compost treatments (Ros et al. 2006). The use of organic fertilisers over chemical fertilisers generally reduces the carbon footprint of farming, since fossil fuels are a key component in the energy-intensive production and subsequent transportation of agrochemicals (Audsley et al. 2009). Furthermore, avoiding the use of chemical fertilisers and pesticides reduces the risk of water and soil pollution. The use of certified organic inputs on organic farms reduces the risk of environmental contamination by chemicals. On some of the organic farms, in addition to promotion of biodiversity indirectly through organic farming practices, direct efforts are made to protect biodiversity and the environment, for example through manual control of invasive alien vegetation, and non-interference with flora and fauna, including reptiles such as snakes. The protection of biodiversity on organic farms is aligned with green economy principles.

4.6   Challenges for Green Economy Implementation and Areas of Misalignment with Green Economy Principles

Some of the practices followed on the farms are not aligned with green economy principles. These can be considered areas of weakness or challenges in terms of meeting green economy imperatives, thus indicating areas that need to be addressed in the process of green economy implementation and opportunities for improvement. These areas of misalignment with green economy imperatives (as summarised in the six green economy themes) are listed in Table 4.5.

---

[4]Microbial biomass (bacteria and fungi) is a measure of the mass of the living component of soil organic matter. The microbial biomass decomposes plant and animal residues and soil organic matter to release carbon dioxide and plant available nutrients.

Table 4.5   Profile of case study farms in terms of green economy weaknesses

Farm	Livelihoods, jobs; addressing inequality and poverty	Governance, participation, agency & capacity	Contribution to economic development	Resource and energy efficiency	Protection of biodiversity and ecosystems
A	• Low skilled labour employed • No equity strategies in place	• Limited opportunities for skills development • Poor water and irrigation management • Expectation of handouts from government	• Cash based—no access to credit • Limited resources for expansion and diversification • Financially self-sustaining	• Routine irrigation—not based on soil condition or crop requirements • Organic fertilisers transported a distance of 500 km to the farm	• Transportation of organic fertilisers over long distances
B	• Low skilled labour employed • No employment for differently abled people • No equity strategies in place	• Expectation of handouts from government • Limited opportunities for skills development • Poor water and irrigation management	• Cash based—no access to credit • No resources for expansion and diversification	• Routine irrigation	• Transportation of organic fertilisers over long distances
C	• Low skilled labour employed • No equity strategies in place • Employee wages below legislated rate	• Poor water and irrigation management • Expectation of handouts from government	• Cash based—no access to credit • No resources for expansion and diversification	• Routine irrigation	• None identified

(continued)

Table 4.5 (continued)

Farm	Livelihoods, jobs; addressing inequality and poverty	Governance, participation, agency & capacity	Contribution to economic development	Resource and energy efficiency	Protection of biodiversity and ecosystems
D	• No female participation in management • No specific efforts to recruit youth • No employment of differently abled people	• Run by male owner—no succession plan • No skills development for all involved • Not inclusive—only female employees	• Unstable marketing arrangements • Full financial benefits not realised—produce is organic but not marketed as organic • Cash based—no access to credit • No resources for expansion and diversification	• Routine irrigation • No control over irrigation time to minimise water loss through evaporation (water only available between 08:00 and 16:00 on weekdays • No composting—crop residues removed and fed to cattle, but manure from the cattle is not used on the farm. • Organic fertilisers transported a distance of 500 km to the farm	• No use of compost or manure
E	• Agrochemical exposure risk for workers • No employment of differently abled people • No female participation in management • No specific efforts to recruit youth	• No skills development for manager & workers • No facilities or opportunities for skills development • Not inclusive—only female employees	• Cash based—no access to credit • No resources for expansion and diversification	• Routine irrigation • Irrigation conducted in morning and afternoon (working hours)—no opportunity for irrigating when water loss due to evaporation is minimal • No composting—crop residues are fed to pigs; but manure from pigs is not used on crops • Fossil fuel use through use of agrochemicals	• Exclusive use of agrochemicals—no integrated pest management • Frequent use of assortment of pesticides and synthetic fertilisers • No compost or manure used • No efforts (both implicit and explicit) to protect the environment

(continued)

Table 4.5 (continued)

Farm	Livelihoods, jobs; addressing inequality and poverty	Governance, participation, agency & capacity	Contribution to economic development	Resource and energy efficiency	Protection of biodiversity and ecosystems
F	• Pesticide risk exposure for workers • No equity strategies in place	• Not inclusive—only female members	• No evidence that enterprise will be self-sustaining—fully funded through government grant	• Routine irrigation • No composting • Exclusive use of agrochemicals—no integrated pest management	• Frequent use of pesticides and synthetic fertilisers • No efforts to protect environment • Elevated carbon footprint through agrochemical use
G	• Pesticide risk exposure for workers • No employment of differently abled people	• No skills development for workers	• Cash based no access to credit • Not financially self-sustaining—subsidised by income from other sources	• Routine irrigation • Exclusive use of agrochemicals—no integrated pest management	• Environment and human risks from frequent use of pesticides and synthetic fertilisers • No efforts (implicit and/or explicit) to protect environment • Raised carbon footprint through agrochemical use • Exclusive use of agrochemicals—no integrated pest management

4.6.1 Livelihoods and Jobs

The jobs provided on the farms cater for unskilled workers. Neither the permanent nor the casual jobs are secure, as binding legal employment contracts are generally not in place. In terms of worker safety, the work involves physically demanding manual labour and long periods in direct sun. Workers on the conventional farms face the risk of exposure to toxic chemicals such as pesticides. In addition, some farmers acknowledged that due to their own financial constraints they are unable to pay their workers the minimum legislated wages, arguing however that "some employment is better than no employment". This situation is not in line with the green economy principles of green jobs and decent work. No additional benefits such as health and retirement contributions are provided for employees either, as farmers cannot afford these benefits. There are no opportunities for employment further along the value chain (e.g. in processing activities) due to the fact that these farms only have the capacity and resources to produce crops.

4.6.2 Governance, Participation, Agency and Capacity

Although there is networking and information exchange through the Nkomamonta Cooperative and a tomato producers' association, there are limited opportunities for on-going capacity building and exposure to up-to-date technical and market information through media such as newsletters, workshops or short courses. Despite running their own farms and thus displaying good agency, most farmers expect to receive government assistance in the form of grants to finance their operations. There is a pronounced sense of entitlement to government support, which at the time of conducting the study was not forthcoming. It is important to note that continuous material assistance from government is not sustainable, and farmers have to develop the capacity to operate independently. They also have to ultimately develop their own information and knowledge sharing networks.

4.6.3 Inequality and Poverty

The predominant employment of women on the farms is not driven by an acknowledgment of gender issues; rather, some farmers note that women were more willing to work as labourers and are better at the work than men. In fact, one could argue that this observation actually supports patriarchal values rather than gender equality. The jobs available on the farms are limited to unskilled manual work and the farms do not employ differently abled individuals (people with disabilities), due to the nature of the work. No specific strategies to ensure equity in employment for specific groups such as youth and women are in place on any of the farms.

4.6.4 Contribution to Economic Development

Although all the farms sell produce, some are not fully utilising the economic potential of their produce. Due to non-renewal of organic certification and unavailability of certified organic produce packing and handling facilities locally, some organic farmers struggle to access niche markets for their produce, and end up not marketing their produce under the organic label. This limits the returns on the investments they make into organic farming. From an economic perspective, some farms are not fully utilising the potential to contribute to the economic development of the area and beyond. One of the reasons for this is that farmers have no access to credit and thus have no financial resources to expand and/or diversify their operations. Farmer F, for example, highlighted that demand for their produce on the local informal market exceeds supply, but the farm could not increase production due to inadequate finance. The farmers' limited financial resources reduce the scale of production and consequently the output and contribution to the economy. In addition, the limited financial resources confine farmers to crop production, and preclude them from diversifying into local-level value addition.

4.6.5 Resource and Energy Efficiency

A core part of water conservation in irrigation is water management through accurate irrigation scheduling. Irrigation scheduling means determining the timing and duration of irrigation and the amounts of water applied; based upon crop needs, soil water storage capacity and climatic conditions; in an effort to ensure more efficient water use. A common weakness across all farms is that irrigation is applied routinely and is neither based on soil water measurements nor crop water requirements. The farms may be applying too much or too little water relative to crop requirements, and thus using water inefficiently.

Most of the conventional farms discard waste such as crop residues and substandard produce; instead of composting it and applying it to the soil to improve soil fertility and soil water retention, which would in turn reduce the levels of other inputs required. Discarding these residues can therefore be seen as a waste of resources. Two farms (D and E) use crop residues to feed livestock. However, the manure from the livestock is not used for crop production; and this can be considered an inefficient way of using crop residues as the manure from the livestock could be used to fertilise crops.

All the conventional farms use agrochemicals, and this indirectly increases their non-renewable energy use, as fossil fuels are used in the production and transportation of agrochemicals. On the other hand, the transportation of organic inputs over long distances by individual organic farmers implies significant use of energy (fossil fuels used in transportation); which could potentially be addressed through bulk transportation by a local distributor or by the farmers coordinating their organic input purchases and sharing transport to increase efficiencies.

4.6.6 Protection of Biodiversity and Ecosystems

For the organic farms, organic inputs and fertilisers are transported over a distance of 500 km, as there are no local suppliers of organic inputs; with each farmer transporting their inputs individually. This is not only costly for the farmer, but implies relatively high carbon emissions per unit of fertiliser. The conventional farms are characterised by frequent application of an assortment of pesticides, especially on tomatoes. This frequent use of pesticides has negative implications for biodiversity and human health. When agrochemicals are applied, they also affect non-target species and habitats. Runoff can carry pesticides into aquatic environments, while wind can carry them to areas outside the sprayed fields, potentially affecting other species. Agricultural pesticides have been linked to invertebrate biodiversity loss (Beketov et al. 2013; Goulson 2013). Commonly used agricultural pesticides sprayed on crops (neonicotinoids) have been linked to the death of bees and even birds (Henry et al. 2012; Goulson 2013).

Most of the conventional farms only use inorganic fertilisers. The use of inorganic fertilisers without manure or organic fertilisers results in a decline in soil organic carbon and soil microorganisms (biodiversity) and water holding capacity (Brussaard et al. 2007). The carbon footprint of the conventional farms is thus raised by the use of agrochemicals, and on these farms there is a risk of both groundwater and surface water pollution. The risks to biodiversity and ecosystems posed by conventional production methods represent constraints for the establishment of green economy initiatives. These factors would have to be addressed if conventional vegetable farms are to contribute to a green economy.

4.7 Implications of Socio-Economic and Biophysical Conditions in Which Farmers Operate for Green Economy Implementation

The biophysical and socio-economic conditions of a farming area have implications for green economy implementation, as these conditions can impede or facilitate implementation. This is illustrated in this section through analysis of the case study farms in the context of the broader conditions in the GTLM. In a crop production context, the climatic and other biophysical conditions of an area are key determinants of the types of crops that can be produced and their productivity. The biophysical environment of the GTLM, characterised by a subtropical climate, is conducive for the production of a variety of high value crops, mainly fruits, nuts and vegetables; and the area has a well-developed agricultural sector (GTLM 2014). This is conducive for agricultural green economy implementation, as the requisite infrastructure and support services such as input supply (except organic inputs), transport and marketing networks would already be in place. Agricultural production and value chain development are strategic focus areas for the development of the economy of the Tzaneen municipality (GTLM 2014, 2018). The prominence of agriculture in the municipality's local economic development plans (GTLM 2014, 2018) presents opportunities for agricultural green economy projects in the area; as such initiatives would be compatible with the local government's plans and thus likely to receive the requisite policy support.

At the broader level, the socio-economic conditions in GTLM present both opportunities and challenges for green economy implementation in the agriculture sector. The fact that the majority of residents live in rural villages (GTLM, 2014, 2018) and have access to land and secure tenure, as well as the availability of water for irrigation, are all factors that are conducive for the development of an agricultural green economy based on irrigated crop production. This is borne out by the fact that all case study farmers have access to land and water. The availability of these resources highlights the fact that the basic natural resources for green economy implementation in agriculture are available in the Tzaneen area.

Infrastructure is critical for agricultural production and marketing. The case study farmers have access to infrastructure such as dams, roads and electricity; which enables them to produce and market crops. At the broader level, the basic infrastructure that would be required for an agricultural green economy, such as roads and electricity, is also available to the majority of residents. However, problems with the maintenance of gravel roads, and limitations with provision of new electricity infrastructure and maintenance of existing infrastructure, as alluded to in GTLM development plans (GTLM 2014, 2018), could hamper green economy implementation in the area.

Social aspects have implications for green economy implementation, as skills and human resources are required to operate green economy projects. Some of the social conditions in GTLM, such as the prevalence of female-headed households and low income levels, could hinder green economy implementation. The responsibilities that women have as household heads may limit availability of human resources to spearhead green economy initiatives. Furthermore, the low income levels in the area could limit the availability of resources for investment in green economy initiatives. Alternatively, the low income levels could also provide motivation for people to try new initiatives to supplement their incomes. Finally, the low education levels could pose potential problems for capacity building for implementation of green economy projects; although that could also encourage people to learn new skills.

Access to credit is an important factor for green economy implementation. The case study farmers have difficulties in accessing credit; which constrains their ability to expand their operations. In an environment such as GTLM, where levels of poverty and unemployment are high, access to financial resources to invest in agricultural green economy initiatives is likely to be difficult. Green economy implementation would thus have to address financing issues. Marketing is another key aspect of green economy development. All the case study farmers produce a variety of vegetables for sale, and thus have linkages to markets and some understanding of operating within an agricultural value chain. This exposure presents opportunities for green economy development; as existing market linkages could be leveraged and the existing capacity and experience of farmers utilised. The fact that farmers already produce a variety of vegetables would be positive for the development of a green economy, as this means that the knowledge to produce different crops already exists; and the risks associated with a single crop, for example unfavourable market or weather conditions, could be minimised.

Relevant and up-to-date knowledge and information are necessary for green economy implementation. An operating environment that has institutions that are appropriate for knowledge and information dissemination would facilitate green economy implementation. In the case of Tzaneen, organisations such as the Nkomamonta Organic Farmers' Cooperative and the Limpopo Tomato Growers Association could be used as platforms for building capacity and disseminating knowledge and information on agricultural green economy practices. The cooperative would also be useful for facilitating organic certification for individual farmers and access to markets, as has been the case in the past. The provincial government agricultural advisory service, LDARD, already provides information and technical support to farmers, and would be an appropriate provider of information and support relevant to the green economy. The mentoring arrangement for new farmers facilitated by LDARD would be another avenue for building green economy implementation capacity.

The accessibility of requisite goods and services is critical for green economy implementation. The local availability of necessary farming inputs has implications for costs (both environmental and monetary) and hence profitability of agricultural ventures, and by extension, for agricultural green economy initiatives. Each of the organic farmers in our study has to transport their own organic inputs over long distances, which is wasteful in terms of resource use and carbon emissions. Bulk transportation of inputs to the area would lower unit costs of inputs (in both monetary and carbon emission terms). This could be done by local input supply businesses or by farmers pooling their resources or through the organic farmers' cooperative.

4.8 Lessons from the Case Studies for Green Economy Implementation

The case studies show that agricultural production methods have implications for meeting green economy imperatives. The organic farms generally had better alignment with green economy environmental principles than the conventional farms. This does not mean that green economy projects have to follow organic production methods; however, the case studies highlight that practices which incorporate environmental protection and human safety are central to green economy implementation.

Although green economy implementation in a crop production context occurs at project or farm level, there are factors that are essential for successful implementation which require an enabling policy environment. For example, organic farmers in Tzaneen cannot access organic inputs locally, and this has negative implications for resource efficiency and carbon emissions. An enabling environment for the green economy would have the necessary policies and incentives to ensure that the whole operating environment facilitates a green economy. In the case of Tzaneen, a policy to incentivise local businesses to stock organic inputs would contribute to an enabling environment for the green economy.

The lessons learned are discussed in more detail under each of the green economy themes below.

4.8.1 Livelihoods and Jobs

One of the lessons from the case studies is that commercially-oriented small-scale farming can provide local employment and livelihood opportunities. Although each farm only employs a relatively small number of manual workers at low wage levels, this type of employment can contribute towards employment creation, especially in rural areas where employment opportunities are few and skill levels are low. People employed on vegetable farms in Tzaneen also have access to free food, as excess produce, as well as produce that does not meet the market grade, is made available to employees. In addition, the multiplier effects of many small-scale projects can contribute towards livelihoods for thousands of people. The value of agricultural green economy projects in employment creation should therefore not be underestimated. As such, green economy implementation in the agriculture sector could contribute to addressing intractable African human well-being issues such as poverty, unemployment and food security.

4.8.2 Governance, Participation, Agency and Capacity

The case studies highlight the involvement of both men and women in farming operations on family farms. On some of the farms, there is intergenerational involvement in the enterprises, indicating some level of planning for succession and continuation of farming operations into the future. A lesson from this situation is that green economy implementation has to be cognisant of prevailing issues of governance, agency and capacity in the local context, and should build on existing arrangements. Green economy projects would for example have to ensure the full involvement of the different role players in family farming to ensure that all involved have the requisite capacity to fully participate in, and make informed decisions on green economy projects, and to carry such projects into the future.

Managing the expectations of those participating in green economy implementation and building self-sustaining initiatives is important. Despite operating for some years, many farmers in the case studies still express a desire for material support from the government. If and when material support is provided to farmers for green economy implementation, this should be done in a way that does not create expectations of, or reliance on continued handouts on the part of the recipients. Continuous material assistance from external entities is neither feasible nor sustainable. Support provided for green economy implementation has to be designed to ensure that farmers are capacitated to sustain their operations independently.

4.8.3   Addressing Inequality and Poverty

Each of the farms employs workers, mainly women, on either a permanent or casual basis. Some also employ young people. However, generally there are no employment equity strategies on any of the farms; and the employment of women or the youth is mainly for other reasons, not addressing inequality. Although the number of employees on each farm is relatively low, the farms are contributing towards poverty reduction, as they provide employment in an area with high levels of poverty and unemployment. The creation of jobs on farms resonates with South Africa's green economy objectives.

There are no differently-abled people employed on any of the farms. The nature of the work (manual labour) makes it impractical for people with disabilities to participate. If vegetable value chains are developed in a green economy, some of the work, for example in packing, would be amenable to the participation of differently-abled people. The employment situation on the farms is such that it is mainly women employed in arduous, low-paid jobs as farm labourers. This could be a reflection of the desperation of women as heads of household in a situation where livelihood options are limited. A green economy, if properly conceived, could open up opportunities for more and better paying jobs. Given the prevailing situation, in addition to other imperatives, a green economy would have to focus on addressing issues of employment creation (in terms of both quantity and quality of jobs).

4.8.4   Contribution to Economic Development

The case studies highlight that small-scale green economy agricultural projects can contribute towards local and national economies. The farms contribute to the local and national economy as they procure inputs from local suppliers and employ workers from the local community (both of which would in turn have multiplier effects), and supply produce to local and national markets. The local market includes sales to informal bulk traders, who in turn support a local vending industry which provides further livelihood opportunities and contributes to the local economy. In areas such as

Tzaneen where the concept and framework for producing for sale exists among small-scale farmers, these can provide a strong base for an agricultural green economy to build on.

Green economy implementation has to provide support to ensure financial viability of enterprises. Such support would have to cover financial planning and management, among other issues. Financial sustainability is an issue in Tzaneen, with only one of the case study farmers perceiving their vegetable enterprise as financially sustainable. The prevailing situation is that returns from sales of produce cannot cover full production and marketing costs; and enterprises are subsidised with funds from other sources, including off-farm employment and retirement pensions. Green economy implementation would also have to ensure that farmers have access to credit in order to finance their operations and to expand and diversify as required.

4.8.5 Resource and Energy Efficiency

One lesson from the case studies is the importance of matching equipment with relevant technical and managerial skills. All the farms use drip irrigation in order to minimise water use. However, the drip irrigation equipment is not matched with irrigation management skills to ensure optimum operation of the irrigation systems and maximum benefits. A common factor across all the farms is that irrigation is applied routinely and not based on soil water measurements or crop water requirements. As such, the water saving benefits of drip irrigation systems are not fully realised. Green economy implementation would have to ensure that farmers are capacitated to manage every aspect of their enterprises efficiently and to use best practice.

Green economy implementation has to be holistic, so that green economy projects are conceived and managed in ways that build synergies with other enterprises on mixed farms and that use resources efficiently. On two farms, crop residues are fed to livestock, but the manure from the livestock is not used to fertilise soils. This represents an inefficient way of using resources, and an unutilised opportunity to build synergies between livestock and vegetable enterprises. Some farmers compost crop residues and other organic materials available to them such as chicken manure, and use the compost to fertilise crops. These farms provide examples of the efficient use of resources that is central to a green economy. Such farms could also provide learning during green economy implementation for farmers who perceive such practices as onerous or ineffective.

4.8.6 Protection of Biodiversity and Ecosystems

One of the lessons from the case studies is that small-scale farmers are willing and able to make investments that benefit both the environment and crops. In particular, organic farms make use of organic matter as compost, and procured organic inputs. Compost builds up soil organic matter, and this in turn promotes soil biodiversity, which has both short and long-term benefits for crops and the environment. Undertaking farming practices that are aligned with the green economy is therefore achievable.

The case studies highlight the environmental issues that green economy implementation has to address in a small-scale crop production context. The farms which follow conventional production methods use pesticides frequently; which can have negative impacts on biodiversity and human health (farm workers and consumers). Most of these farms only use inorganic fertilisers; which can result in a decline in soil organic carbon, soil microorganisms (biodiversity) and water holding capacity. The carbon footprint of these farms is raised by the use of agrochemicals, as the production of agrochemicals requires fossil fuels. There is also a risk of both groundwater and surface water pollution with the use of chemical fertilisers and pesticides on these farms. Green economy implementation would have to balance adequate crop production with environmental protection in order to protect livelihoods, human health and the environment. It would also have to proffer alternatives to practices that pose risks to the environment. The alternatives would have to be acceptable to farmers and workable within their operating contexts.

4.9 Conclusion

The concept of a green economy is not a 'one size fits all' and its execution should be tailored for specific contexts. This requires a good understanding of the farming context into which green economy implementation is to occur; including the areas of alignment/misalignment between a farming system and green economy principles. The chapter examines on-farm realities in green economy implementation and is informed by analyses of small-scale vegetable production enterprises in South Africa. An understanding of on-farm factors is important for formulating appropriate green economy implementation interventions. The farms studied had different attributes; for example, some used organic production methods that were more aligned with green economy principles; while others used conventional production methods; most of which did not align with environmental green economy ideals. The key lesson is that green economy implementation should be based on practical realities on the ground and be a process of building onto compatible practices where these occur; and introducing alternatives to practices which are not compatible with green economy principles.

Acknowledgements We thank the Limpopo Department of Agriculture and Rural Development for facilitating the case studies and the farmers in Tzaneen for their participation. The contributions of Elliot Moyo and Benita de Wet to the field work are acknowledged.

References

Allen C (2012) A guidebook to the green economy. Issue 2: exploring green economy principles. United nations department of economic and social affairs (UNDESA): United Nations Division for Sustainable Development

Aliber M, Hall R (2012) Support for smallholder farmers in South Africa: challenges of scale and strategy. Dev South Afr 29:548–562

Aliber M, Hart T (2009) Should subsistence farming be supported as a strategy to address rural food security? Agrekon 48:434–458

Audsley E, Stacey K, Parsons D, Williams AG (2009) Estimation of the greenhouse gas emissions from agricultural pesticide manufacture and use. https://dspace.lib.cranfield.ac.uk/bitstream/1826/3913/1/Estimation_of_the_greenhouse_gasemissions_from_agricultural_pesticide_manufacture_and_use-2009.pdf. Accessed 15 Sept 2018

Aujla MS, Thind HS, Buttar GS (2005) Cotton yield and water use efficiency at various levels of water and N through drip irrigation under two methods of planting. Agric Water Manag 71:167–179

Beketov MA, Kefford BJ, Schäfer RB, Liess M (2013) Pesticides reduce regional biodiversity of stream invertebrates. PNAS 110:11039–11043

Brussaard L, de Ruiter PC, Brown, GC (2007) Soil biodiversity for agricultural sustainability. Agric Ecosyst Environ 121:233–244

CIAT (2009) CIAT's medium-term plan 2010–2012. Cali, Colombia. https://cgspace.cgiar.org/bitstream/handle/10568/54626/mtp_2010_2012_jun09_abridged_version1.pdf?sequence=33&isAllowed=y. Accessed 28 Sept 2018

Chikazunga D, Paradza G (2013) Smallholder farming: a panacea for employment creation and enterprise development in South Africa? Lessons from the pro-poor value chain governance project in Limpopo Province. PLAAS working paper, vol 27. https://www.plaas.org.za/publication-categories/wp. Accessed 15 Sept 2018

DEA (2007) Enviropeadia: South Africa's green economy strategy, compiled by chief directorate: communication, department of environmental affairs, Pretoria. http://www.enviropaedia.com/topic/default.php?topic_id=342. Accessed 28 Sept 2018

DEA (2011) National strategy for sustainable development and action plan (NSSD 1). 2011–2014

DPME (2014) Medium term strategic framework 2014–2019. https://www.gov.za/sites/default/files/MTSF_2014-2019.pdf. Accessed 21 Aug 2018

EDD (2011) New growth path: accord 4. Green economy accord. http://www.economic.gov.za/communications/publications/green-economy-accord. Accessed 25 June 2014

FAO (2001) Irrigation water management: Irrigation methods. Manual No 5. ftp://ftp.fao.org/agl/aglw/fwm/Manual5.pdf. Accessed 06 Nov 2014

FAO (2012a) FAO@Rio+20: Greening the economy with agriculture (GEA) —Taking stock of potential, options and prospective challenges. Concept Note http://www.uncsd2012.org/content/documents/GEA__concept_note_3March_references_01.pdf. Accessed 15 Aug 2013

FAO (2012b) Smallholders and family farmers. http://www.fao.org/fileadmin/templates/nr/sustainability_pathways/docs/Factsheet_SMALLHOLDERS.pdf. Accessed 21 Aug 2018

Farming First (2018) Agriculture for a green economy: improved rural livelihood, reduced footprint, secure food supply http://www.farmingfirst.org/wordpress/wp-content/uploads/2011/10/Farming-First-Policy-Paper_Green-Economy.pdf Accessed 19 Sept 2018

Goulson D (2013) Review: an overview of the environmental risks posed by neonicotinoid insecticides. J Appl Ecol 50:977–987

GTLM (2014). Integrated development plan 2014/15. Final http://www.tzaneen.gov.za/tzaneen/GTM%20Final%20Approved%20IDP%202014-2015%20Review%20(2).pdf. Accessed 15 Aug 2018

GTLM (2018) Integrated development plan 2018/19. http://www.greatertzaneen.gov.za/documents/idp/Final%20Approved%20IDP%202018-19.pdf. Accessed 15 Aug 2018

Henry M, Béguin M, Requier F, Rollin O, Odoux JF, Aupinel P, Aptel J, Tchamitchian S, Decourtye A (2012) A common pesticide decreases foraging success and survival in honey bees. Science 336:348–350

IDC, SA LED Network (2011) Development agencies in practice: GTEDA revitalising local economic potentials

IFAD (2013) Smallholders, food security, and the environment. http://www.ifad.org/climate/resources/smallholders_report.pdf. Accessed 23 Jan 2015

Jackson T, Victor PA (2013) Green economy at community scale. George cedric metcalf charitable foundation, Canada. http://metcalffoundation.com/wp-content/uploads/2013/10/GreenEconomy.pdf. Accessed 27 Sept 2018

Mateo N, Ortiz R (2013) Resource efficiency revisited. In: Hershey CH, Neate P (eds) Eco-efficiency: from vision to reality (Issues in Tropical Agriculture series) Cali, CO: Centro Internacional de Agricultura Tropical (CIAT), CIAT Publication No. 381, pp 1–19

McIntyre BD, Herren H, Wakhungu J, Watson RT (eds) (2009) Agriculture at a crossroads, synthesis report: a synthesis of the global and sub-global IAASTD reports. IIAASTD. Island Press, Washington

NPC (2011) National development plan. http://www.npconline.co.za/medialib/downloads/home/NPC%20National%20Development%20Plan%20Vision%202030%20-lo-res.pdf. Accessed 27 Sept 2018

Pascual JA, Hernandez T, Ayuso M, Garcia C (1997) Changes in the microbial activity of arid soils amended with urban organic wastes. Biol Fertil Soils 24:429–434

Randriamaro Z (2012) Greening the economy and increasing economic equity for women farmers in madagascar. International research policy brief, vol 34. The International Policy Centre for Inclusive Growth

Ros M, Knapp KB, Aichberger K, Insam H (2006) Long-term effects of compost amendment of soil on functional and structural diversity and microbial activity. Soil Use Manag 22:209–218

Soil Conservation Council of Canada (2001) http://www.soilcc.ca/downloads/factsheets/Factsheet%203%20-fossil%20fuel.pdf. Accessed 27 Sept 2018

StatsSA (2011) Census 2011, http://beta2.statssa.gov.za/. Accessed 27 Sept 2018

StatsSA (2018) Statistical release P0211- quarterly labour force survey. Quarter 2 2018. http://www.statssa.gov.za/publications/P0211/P02112ndQuarter2018.pdf. Accessed 19 Aug 2018

Tzaneen Information (2013) Discover Tzaneen, vol 2. http://www.tzaneeninfo.co.za/DTM/2013/DiscoverTzaneenMagazine2013.pdf. Accessed 27 Sept 2018

UNDP (2012) Comparative experience: examples of inclusive green economy approaches in UNDP's support to countries. http://www.undp.org/content/dam/undp/library/Environment%20and%20Energy/Examples-of-Inclusive-Green-Economy-Approaches-in-UNDP%27s-Support-to-Countries-June2012_Updated-Sept2012.pdf. Accessed 27 Sept 2018

UNECA (2012) A green economy in the context of sustainable development and poverty eradication: what are the implications for Africa? http://www1.uneca.org/Portals/rio20/documents/cfssd7/1AfricaGE-BackgroundReportEN.pdf Accessed 15 Sept 2013

UNEP (2014) Green economy success stories: organic agriculture in Cuba. http://www.unep.org/greeneconomy/SuccessStories/OrganicAgricultureinCuba/tabid/29890/Default.aspx. Accessed 18 Dec 2014

UNEP (2016) Trade in certified organic agriculture – challenges and opportunities for South Africa. UNEP, Geneva. http://www.un-page.org/files/public/unep_south_africa_2016_76pp1.pdf Accessed 19 Sept 2018

University of Massachusetts (2014) An overview of drip irrigation. https://extension.umass.edu/vegetable/articles/overview-drip-irrigation. Accessed 6 Nov 2014

World Bank (2018) Overcoming poverty and inequality in South Africa: an assessment of drivers, constraints and opportunities. http://documents.worldbank.org/curated/en/530481521735906534/pdf/124521-REV-OUO-South-Africa-Poverty-and-Inequality-Assessment-Report-2018-FINAL-WEB.pdf. Accessed 19 Aug 2018

World Farmers' Organisation (2012) Agriculture's contribution to the green economy: proposed outcomes from the Rio +20 summit. Italy, Rome

Zhen Z, Liu H, Wang N, Guo L, Meng J, Ding N, Guanglei W, Gaoming J (2014) Effects of manure compost application on soil microbial community diversity and soil microenvironments in a temperate cropland in china. PLoS ONE 10:e108555. https://doi.org/10.1371/journal.pone.0108555. Accessed 19 Aug 2018

Chapter 5
Moving from Theory to Practice: A Framework for Green Economy Project Implementation

5.1 Context: Organising and Integrating Information Systematically

While the potential benefits of a green economy are well articulated (e.g. OECD 2011a; UNEP 2011; UNECA 2012), the green economy concept largely exists in the theoretical realm. For green economy aspirations to be realised, the concept has to be implemented at project level. A project can be defined as "a series of activities aimed at bringing about clearly specified objectives within a defined time-period and with a defined budget" (European Commission 2004). In the particular context of this book, the term project is used to refer to activities focused on production of a specific crop; for example producing a tomato crop would be referred to as a tomato production project. A farming enterprise comprised of different crops and/or livestock can be considered a project. However, for purposes of illustrating the various issues involved in green economy project implementation, the term will be used in the narrow context of a specific crop.

The green economy literature currently lacks practical, hands-on, sector-specific tools and techniques to inform implementation at project level. A general literature search on green economy implementation guidelines, methodologies or processes conducted in February 2018 did not yield any results in terms of information of relevance at the project level. Rather, the available literature on green economy implementation focuses on what is required to create an enabling environment for a green economy at country or sector level, highlighting factors such as policies, institutions, opportunities and challenges; including the theoretical, methodological and policy advances that are needed to build a green economy (e.g. Manhong et al. 2011; Richardson 2013; Smith et al. 2014).

Putting theoretical concepts such as the green economy or sustainable development into practice can be challenging (Steelman et al. 2015). This requires moving beyond principles, and effecting substantive action at the local level (Bellamy and Johnson 2000; Bellamy et al. 1999). Amaruzaman et al. (2017) report on what they

C. Musvoto et al., *Green Economy Implementation in the Agriculture Sector*, SpringerBriefs in Agriculture, https://doi.org/10.1007/978-3-030-01809-2_5

term "the performance gap of agriculture in a green economy", alluding to situations in which green economy aspirations have not been matched with action on the ground. Implementation is particularly difficult for a concept such as the green economy; given its multi-faceted nature, with several different definitions and a wide range of principles. In the context of agriculture, there are many diverse factors that are relevant to a green economy, as highlighted in Chaps. 1–4 of this book, all of which have to be considered and acted upon in project implementation. Difficulties with integrating all these factors would negatively affect implementation. According to Amaruzaman et al. (2017), the performance gap of agriculture in a green economy indicates a failure to integrate agricultural development, biodiversity conservation, ecosystem service provision and socially responsible practices.

A key challenge is integrating these different factors in a systematic way to ensure that a project meets both agricultural objectives and green economy imperatives. This is especially important given that agriculture, depending on how it is practised, can have negative impacts on the environment, which could place it at odds with the green economy. A step-by-step process for implementing a project in a way that addresses the potential incompatibilities between the environmental impacts of agriculture and green economy principles, while meeting the socio-economic aspirations of a green economy, is required. This chapter presents such a process in the form of a practical framework, illustrated using the case of vegetable production. The framework is suitable both for new projects, as well as for existing projects that aim to adopt green economy principles. The practical implementation of the framework is also illustrated. It provides step-by-step guidance, from the synthesis of general considerations and factors, to actions at the project level, critical success factors, and actions required to achieve success.

5.2 Identifying Relevant Factors to Be Considered in Project Implementation

Implementing a green economy project (either a new project, or an existing project that will be modified to meet green economy principles) entails considering and translating many factors into coordinated actions. From the desktop review and case studies described in Chaps. 1–4, it is evident that many diverse factors should be considered and addressed in implementing a green economy project. These factors range from generic and contextual factors, to project-level factors. These factors are summarised in Fig. 5.1 and discussed in more detail below.

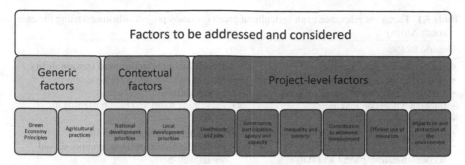

Fig. 5.1 Factors of relevance to green economy project implementation

5.2.1 Generic Factors

There are two categories of generic factors. The first comprises green economy principles, as described in various reports (e.g. OECD 2011a; Allen 2012) and as summarised in Chap. 1, as well as Table 5.1. These principles are the foundation of the green economy concept, and should provide the overarching context for green economy implementation. The second category of generic factors comprises agricultural production and sustainability issues. Green economy implementation in an agricultural context should align the objectives and practices of agriculture with green economy principles. This may, however, not always be straight forward due to the impacts of agriculture on the environment (discussed in Chap. 3). The sustainability of agriculture is therefore central to the alignment of agriculture with green economy ideals.

Agricultural sustainability entails both resilience (the capacity of systems to withstand shocks and stresses) and persistence (the capacity of systems to continue over long periods); and the attainment of multiple outcomes (economic, social and environmental) (Pretty et al. 2008). Components of agricultural sustainability also include maintaining production potential, environmental stewardship, economic viability and social justice (Kirchmann and Thorvaldsson 2000). Related to the green economy notion is the concept of green agriculture, which is characterised by resource efficiency and waste minimisation, together with practices which simultaneously preserve and increase farm productivity and profitability, while maintaining and building ecological resources such as soil and water (UNEP 2011). The World Farmers Organisation (2012) notes four primary goals for agriculture in a green economy: produce more with less, use a knowledge-based approach of best practices, reward farmers for adopting sustainable practices, and break the poverty cycle. All these generic factors have to be considered in green economy implementation.

Table 5.1 Factors of relevance to an agricultural green economy project (illustrated using the case of South Africa)

Generic factors	
Green economy principles • Achieving sustainable development; • Creating decent work and green jobs; • Integrated decision making; • Equity, fairness and justice • Improving governance and the rule of law. • Inclusive; democratic; participatory, accountable, transparent and stable; • Internalising externalities. • Low carbon and energy efficiency • Resource efficiency; • Respect planetary boundaries and ecological limits or scarcity • Protect biodiversity and ecosystems Source: Allen (2012)	**Agricultural production and sustainability** • Meeting human needs for food and other commodities • Preventing harm and health hazards to humans • Profitability/economic viability • Protecting and improving livelihoods and social well-being • Enhancing resilience of people and communities • Promotion of good governance • Protection and enhancement of natural resources • Efficient resource use • Pollution prevention • Ecosystem resilience enhancement • Promotion of good environmental governance • Addressing poverty • Use of best practices • Management of natural resources to enhance present and future production Sources: Kirchmann and Thorvaldsson (2000); Pretty et al. (2008); World Farmers Organisation (2012)
Contextual factors	
At national level– South Africa's green economy should: • Be just, ethical, sustainable • Contribute to global solidarity • Address poverty, unemployment, inequality • Meet basic human needs • Promote youth employment • Promote skills development • Promote decent work • Be based on full cost accounting • Decouple economic growth from resource use and environmental degradation • Be environmentally sustainable • Be climate change resilient • Be low carbon • Be protective of ecosystems • Promote/facilitate sustainable use of natural resources • Protect the environment • Promote/facilitate waste recycling Sources: DEA (2011); EDD (2011); NPC (2011)	At local level—South Africa's green economy should contribute to local development through: • Creating green and decent jobs • Promoting local industry development • Promoting local leadership and ownership, community participation and joint decision-making • Use of local resources and skills Sources: Bond (2002); Koma (2012)

(continued)

Table 5.1 (continued)

Generic factors	
Project-level factors	
• Productivity and profitability • Access to input and output markets • Adequate finance • Creating livelihood opportunities • Contribution to the economy and development (local and national) • Contribution to addressing poverty • Inclusivity and participation • Networking and capacity building • Governance	• Use of resource efficient methods and technology • Appropriate resource management • Suitability of practices for socio-economic environment • Environmental management including climate change mitigation and adaptation • Efficient use of resources and waste minimisation • Protection of biodiversity and natural resources • Crop production practices and suitability for biophysical environment • Minimising carbon emissions

5.2.2 Contextual Factors

A green economy project has to be relevant, and thus has to address real issues within a specific context. These types of issues are illustrated here using the example of South Africa. In South Africa, several government strategies (at both national and local level) highlight the green economy as a vehicle for achieving sustainable development and addressing social issues such as poverty and inequality. An underlying principle of development policies in South Africa is addressing the injustices of past discriminatory policies and practices. This is reflected in the country's key policies and strategies, such as the National Strategy for Sustainable Development (NSSD) (DEA 2011), the National Development Plan (NDP) (NPC 2011), and the Green Economy Accord (EDD 2011). The fundamentals of these strategies have been highlighted in Chap. 4.

Local Economic Development (LED) is at the core of South Africa's development agenda. LED can be defined as a "locally-owned, participatory development process undertaken within a given territory or local administrative area in partnership with both public and private stakeholders" (ILO 2010). In South Africa, LED is understood to be a multi-dimensional and multi-sector process through which the skills, resources and ideas of local stakeholders are combined to stimulate local economies so as to contribute towards job creation, poverty alleviation and the redistribution of wealth (Koma 2012). Green economy projects thus have to align with the specific contextual objectives of the locations in which they are implemented.

5.2.3   Project-Level Factors

Agricultural projects are set up to achieve agricultural objectives; therefore green
economy project implementation has to consider project or farm-level issues. Based
on lessons from the case studies described in Chap. 4, these aspects include production methods (for example following organic vs inorganic practices); and contribution
to social well-being, such as generation of livelihood and employment opportunities,
contribution to local poverty reduction, and providing knowledge and capacity building opportunities for the people involved. These factors are summarised in Table 5.1.

5.3   Translating Relevant Factors into Actions at Project
Level

Given the number and range of pertinent factors that have to be considered (see
Table 5.1), it would be cumbersome to try and consider each of these factors in turn
during the process of implementing a project. It is thus necessary to consolidate
these issues into a condensed (more manageable) set of considerations which would
be easier to interpret and apply at the project level. The resulting condensed set of
considerations would, however, have to capture the essence of all of the relevant
factors, as these are critical for meeting agricultural and green economy imperatives.

The issues listed in Table 5.1 can be consolidated along thematic lines into a
handful of key considerations for project implementation, listed in Table 5.2. The
number of considerations can vary from project to project; but it should be kept as low
as possible so as to keep the information easy to manage. In this case, the fifty-plus
factors listed in Table 5.1 can be distilled into eight considerations (Table 5.2), each
of which incorporates generic, contextual and project-level factors. The job creation
consideration for example, integrates generic factors such as livelihood protection
and decent work; contextual factors such as youth employment and decent work; and
project level factors such as actual number of people employed.

These considerations can be further integrated into green economy project standards. These standards are key criteria that a green economy project should meet
(Table 5.3). While in this case the eight considerations were integrated into five standards, the number of standards would vary from project to project. These standards
are discussed in more detail below.

5.3.1   Standard 1: Low Carbon and Environmental Protection

It is imperative to ensure that agriculture is aligned with green economy ideals around
reducing greenhouse gas emissions and ensuring environmental protection. The agriculture sector is responsible for greenhouse gas emissions, and while estimates vary,

Table 5.2 Integration of issues of relevance into key project implementation considerations

Generic factors	Contextual factors	Project level factors	Consolidated considerations
Factors integrated into considerations →			
• Decent work and green jobs; • Employment creation; • Protecting and improving livelihoods	• Addressing unemployment; • Promoting green and decent work, youth employment	• Number of people employed and nature of employment	Job creation
• Justice, fairness, equity, inclusivity and participation; • Governance and rule of law; • Democracy and accountability; • Improving social well-being; • Promoting good governance; • Enhancing resilience of people and communities	• Addressing poverty, inequality, exclusion; • Meeting basic human needs, justice	• Inclusivity; • Participation; • Networking; • Capacity building; • Governance	Social equity and inclusivity
• Internalising externalities; • Enhancing resilience of people and communities; • Profitability/economic viability	• Promoting local industry development; • Use local resources and skills	• Contribution to economy and development at different levels	Contribution to and supporting development at different levels
• Resource and energy efficiency; • Use of resource efficient methods and technologies	• Sustainable natural resource use; • Decouple economic growth from resource use; • Ecosystem protection; • Environmental sustainability	• Efficient use of resources; • Waste minimisation	Resource efficiency

(continued)

Table 5.2 (continued)

Generic factors	Contextual factors	Project level factors	Consolidated considerations
• Low-carbon; • Respect planetary boundaries and ecological limits; • Protection of biodiversity and ecosystems; • Protection and enhancement of natural resources • Prevention of pollution • Good environmental governance	• Climate resilience; • Low carbon development	• Minimising carbon emissions; • Protection of natural resources	Protecting the environment
• Meeting human needs for food and other commodities; • Livelihood protection and improvement • Preventing harm and health hazards to humans; • Managing natural resources to enhance both present and future production;		• Choice of production practices	Production practices
• Internalising externalities; • Meeting production objectives; • Maintaining profitability/economic viability; • Adequate financing; • Crop choice and suitability for socio-economic and biophysical environments		• Crop choices; • Cropping cycles; • Resource and production efficiency; • General management	Productivity and profitability
		• Input and output market location and access	Marketing

Table 5.3 Integrating green economy project implementation considerations into green economy project standards		Considerations (from Table 5.2)	Green economy project standard
	1	Protecting the environment	Low carbon and environmental protection
	2	Resource efficiency	Resource efficiency
	3	• Job creation • Social equity and inclusivity	Social equity and inclusivity
	4	• Production practices • Productivity and profitability	Sustainability and long term economic viability
	5	• Contribution to and supporting development at different levels • Marketing	Relevance to the local context

the sector is responsible for up to 29% of global anthropogenic greenhouse gas emissions (Vermeulen et al. 2012; CGIAR 2014). This is at variance with the ideals of a green economy. In project implementation, it is essential to identify potential sources of greenhouse gas emissions in a project, and to take measures to minimize them. In addition, projects should be cognisant of and minimise the other environmental risks that are associated with the practice of agriculture; such as pollution from fertilisers and pesticides, the degradation of soils through erosion, and the depletion of soil nutrients and soil carbon.

5.3.2 Standard 2: Resource Efficiency

An agricultural green economy project should use resources efficiently. These include water, energy, and inputs such as fertilisers. Linked to resource efficiency is waste minimisation; and a project should strive to minimize waste of resources, including ensuring that loss of produce through factors such as inappropriate handling and storage and damage by pests is minimised.

5.3.3 Standard 3: Social Equity and Inclusivity

The green economy has a strong social focus, and in many developing countries, including South Africa, social objectives such as creating jobs, reducing poverty and achieving social equity are green economy priorities.

5.3.4 Standard 4: Sustainability and Long Term Economic Viability

An agricultural green economy project has to be able to satisfy the human needs for which it is set up, such as meeting production targets on a sustained basis; and it should also be economically viable. This standard is critical, as it underpins the well-being of farmers and others who derive a livelihood from a project, and is linked to social equity and inclusivity.

5.3.5 Standard 5: Relevance to the Local Context

The green economy is meant to address the various components of human well-being in tangible ways. Green economy projects should therefore not be inward looking, but should strive to be relevant to the local context, and contribute to addressing some of the needs of communities in the areas in which they are located. In South Africa, for example, rural development and LED are highlighted as priorities in government development plans. Green economy projects located in rural areas would therefore have to contribute towards rural development.

In Sects. 5.1–5.3, we have identified the factors that have to be considered in green economy project implementation and integrated them into green economy standards for application at project level. In Sects. 5.4 and 5.5, we explain how the green economy standards identified above can be applied to green economy implementation in the case of existing and new projects, respectively.

5.4 Modifying Existing Projects to Attain Green Economy Ideals

Agriculture is a well-established sector with entrenched methods and practices that have enabled the production of food and other commodities for thousands of years. However, these methods and practices are not always in harmony with the green economy. As such, green economy implementation in the agriculture sector will have to entail making some changes to existing projects and practices. Such changes have to be logical as well as contextually relevant if they are to be acceptable to practitioners. Given the number and diversity of issues that have to be considered and acted upon, modifying ongoing initiatives to meet green economy ideals should be methodical and consistent. Such modification should not unduly disrupt production, nor threaten profitability and thus the sustainability of existing projects. In this context, an existing project is defined as a current system of growing a specific crop that is followed on a particular farm.

Before any changes are effected to an ongoing project, it is critical to fully understand the project. This understanding can be built through first analysing the project, and then using the output of the analysis to inform action to align the project with green economy imperatives. This analysis can be done in a four-step process illustrated here using the case of crop production. Although crop production is used as an example, the process can be applied to any agricultural project. The information presented in this example is not all-inclusive, but provides the key pointers for evaluating an existing project to ensure that key issues for green economy implementation are not missed.

5.4.1 Step 1: Describe the Project in Its Current State

The first step in modifying a project for the green economy focuses on gaining a full understanding of the project. A useful approach for fully understanding a project is to follow its value chain. A generic outline of an agricultural value chain is provided in Fig. 5.2.

Ideally, each step in the value chain should be examined in the process of describing a project. However, since in many cases those involved in one aspect of an agricultural project do not have any involvement in other aspects; it is not generally possible for those involved in a project to influence all stages in the value chain. For example, farmers generally have very little control over what happens to their produce once it leaves the farm and moves up the value chain (Fig. 5.2).

For those who are not directly involved in farming, but are involved in green economy implementation, for example agricultural advisers; understanding of a project would be built through directly collecting information on the project. This could be done through speaking to the key people involved (e.g. the farmer) to gather the relevant information. Key information that is necessary for understanding a project would include the specific context of the project, as well as information on the green economy standards, such as those identified in Sect. 5.3 above. Typical issues that should be covered in the description of a project are listed in Table 5.4, categorised according to the green economy standards identified in Sect. 5.3.

First-hand information gathering through observation is also essential for building understanding of a project. Observation occurs at the site of a project in order to verify practices and information gathered indirectly through interviews. Observation should focus on details such as crops growing and their general state; the presence of workers; the equipment in use, such as the type of irrigation system, pumping equipment; and sources of power and/or fuel. Environmental issues such as soil erosion, pollution, recycling, conservation structures and other visible impacts on the environment should also be covered by the observation. The issues listed in Table 5.4 would be a useful guide for project observation.

Fig. 5.2 Use of value chain analysis in building understanding of an ongoing project

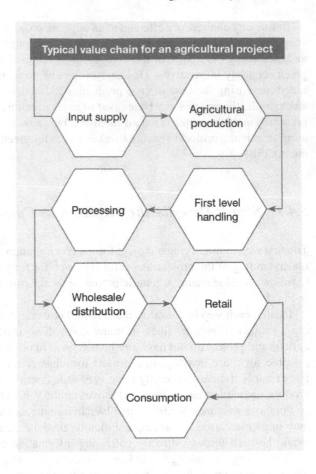

5.4.2 Step 2: Project Screening

After building the general profile of a project, the second step in modifying a project for the green economy is the screening of a project to establish the extent to which it meets the green economy standards identified above. The screening process is based on a good understanding of the role of agriculture in a green economy as provided in the preceding chapters of this book, and the issues that are embedded in each of the green economy standards (as described in Sect. 5.3). Relevant questions would have to be asked about a project against each of these standards; such as the extent to which it meets the standards associated with social equity and inclusivity. Illustrative project screening questions and answers are listed in Table 5.5.

Table 5.4 Indicative issues to cover in the process of describing a project

Context and project level green economy standards	Specific Issues
Specific context	• Crops grown/to be grown • Production objectives: subsistence and/or commercial • Any environmental objectives • Any social objectives • Current problems (for existing projects) • Envisaged problems (for planned projects)
Low carbon and environmental protection	• Production methods and practices • Inputs used/to be used and source • Pest control methods used/to be used • Energy types and sources • Measures to protect the environment e.g. erosion control
Resource efficiency	• Use of irrigation; including type • What does/will inform irrigation decisions? • Use of compost/manure/organic soil amendments • Resource conservation measures e.g. water conservation, energy conservation
Social equity and inclusivity	• Employment of people • Employment of women and youth • Capacity building for those involved in project
Sustainability and long term economic viability	• Source of finance for the project • Markets—location and means of accessing the markets • Value addition to produce at farm level • Is/will the project be financially self-sustaining?
Relevance to the local context	• Sale of produce to local markets • Employment of local people • Procurement of goods and/or services from local markets

5.4.3 Step 3: Green Economy Strength and Weakness Identification

This step identifies a project's strengths and weaknesses in relation to the green economy standards, based on the screening process outlined above. Areas of alignment with green economy standards (i.e. where the standards are being met) are considered as strengths; while areas of misalignment (standards not being met) are considered as weaknesses (Table 5.5).

Table 5.5 Illustrative screening and determination of green economy strengths and weaknesses in an ongoing agricultural project

Evaluation questions based on green economy standards	Relevant project information	Green economy alignment
Does the project address social equity and inclusivity; and how?	Employs 3 women	Strong
	No employees	Weak
	All employees have employment contracts	Strong
	Employment agreements are verbal and negotiable at any time	Weak
Does the project use resources efficiently?	Use drip irrigation to save water	Strong
	No irrigation between 10:00 and 16:00 to minimize water loss through evaporation	Strong
	Irrigation is carried out whenever it is convenient to do so	Weak
Does the project have in place measures for reducing carbon emissions and protecting the environment; what measures?	Transport is pooled with neighbouring farmers (for transporting produce to market and input procurement)	Strong
	Trips to market and to procure inputs are made whenever required –there is no advance planning	Weak
	Pesticides are only used when pest densities are above certain thresholds	Strong
	Pesticides are applied routinely as a preventative measure	Weak
Is the project relevant to the local context and how?	Inputs and services are routinely procured from the local community and if not available from the nearest town	Strong
	All produce is sold through formal contracts to supermarket chains in the capital city; none is sold to local markets—transacting with local markets is time consuming	Weak
	Second grade produce is sold to local traders for resale to local consumers	Strong
Are there measures in place to ensure the project's sustainability and long term economic viability	Staff receive training and information on climate change adaptation	Strong
	Records of income and expenditure are kept and analysed regularly; and high expenditure is curtailed	Strong
	Financial record keeping is time consuming. Income is used as required to meet expenses	Weak

5.4.4 *Step 4: Adjustments for Green Economy Alignment*

After identifying a project's green economy strengths and weaknesses, the next step is to identify the adjustments that are necessary to better align a project with green economy standards; that is, building on strengths and addressing weaknesses. Effecting these adjustments is the process of implementation; which is described in Sect. 5.6.

5.5 Conceptualising a New Green Economy Project

In contrast to Sect. 5.4, where we considered modifying an existing project to meet green economy standards; this section touches on the case of new agricultural projects. As the project is conceptualised, it should be aligned with green economy ideals from the outset. An appropriate starting point is the general aims of the project. These aims should be holistic and cover different aspects, such as agricultural production, as well as environmental, economic and social aspects. The conceptualisation should cover a project's context, as well as project-level green economy standards (as per Sect. 5.4); specifying factors such as the crops that will be grown and the production methods that will be followed. The issues listed in Table 5.4 provide a guide for project conceptualisation. Conceptualisation is meant to produce a rough blueprint of the project; with the details being worked out in the step-by-step project implementation process (Sect. 5.6).

Conceptualisation should as far as possible consider the whole value chain of the project, as discussed in Sect. 5.3. However, due to the fact that in most instances, a specific actor (e.g. a farmer) will only have complete control over one part of the value chain (e.g. crop production), it may be impossible to cover every part of a project's value chain during conceptualisation.

5.6 A Step by Step Project Implementation Process

Implementation is the process of carrying out or putting a decision or plan into effect. The project implementation process described here applies to both new projects and existing projects that are being modified for the green economy.

5.6.1 *Step 1: Characterising a Project*

Once a rough outline of a new project has been built and an existing project is fully understood (Sects. 5.4 and 5.5), the next step is to define a project's objectives in a green economy context. This would be informed by each of the project level green economy standards (defined in Sect. 5.3). Since a green economy project has to address social, economic and environmental issues, it is expected that such a project would have more than one objective. In the case of modifying existing projects for alignment with green economy ideals, the objectives would be focussed around improving those aspects of a project that are currently misaligned with green economy principles (or project standards), and/or building onto areas that are already aligned. Typical objectives for such a project would centre on adjusting current practices to better align with green economy ideals, such as optimising the use of resources like water and energy; and minimising risks to the environment, such as fertiliser leaching. Where a new project is being designed, the focus of the objectives would be on setting up the right conditions for meeting green economy ideals from the outset; such as using practices that minimise risks to the environment. Examples of possible objectives for green economy projects are provided in Table 5.6.

In addition to spelling out a project's objectives, the process of characterising a project should also describe what a 'successful' project would look like. Success can be defined in terms of Critical Success Factors (CSFs); that is, those aspects that are essential for a project to meet its objectives. In a green economy project, CSFs are the areas in which good performance is essential to ensure attainment of green economy ideals and meeting of agricultural objectives. An example of a CSF for a green economy project could be prevention of environmental degradation through control of factors such as soil erosion, leaching of agri-chemicals, indiscriminate pesticide use, etc. Examples of CSFs for green economy vegetable production projects (both new and existing) are listed in Table 5.6. These CSFs are context-specific and can be used to further refine the objectives to enhance alignment with green economy ideals. CSFs also inform project implementation and monitoring actions.

5.6.2 Step 2: Identifying Actions Required to Achieve Success

Once the critical success factors are defined, the next step is to identify the specific actions that should be taken under each objective to achieve the CSFs. These actions would be informed by the specific context of a project. In South Africa, a green economy project would be expected to contribute towards addressing poverty and inequality. A project might make this contribution by employing people, and specifically targeting groups such as women and the youth, who generally experience higher rates of unemployment in comparison with older males. Actions to create employment opportunities would include minimising mechanisation and using labour-based methods. In the creation of employment opportunities, a green economy project would have to take action to ensure that the jobs created align with other green economy ideals, for example ensuring that the jobs created are safe, i.e. do not harm people or the environment.

In the South African context, the green economy is expected to enhance rural development. Given the limited economic opportunities in rural areas and the fact that agricultural green economy projects have a high likelihood of being located in

Table 5.6 Characterising a green economy project

Project standard	Objectives for modifying an existing project	Critical success factors for modifying an existing project	Objectives for design of a new project	Critical success factors for a new project
Low carbon and environmental protection	Reduction and eventual elimination of practices that damage soil health (soil physical, chemical and biological properties) and adoption of those that promote soil health properties)	Improvement in soil health	Use of practices that promote soil health (enhance soil physical, chemical and biological properties	Healthy soil
	Minimise negative impacts on the environment	Avoiding environmental degradation and conserving natural resources	Minimise negative impacts on the environment	Avoiding environmental degradation and conserving natural resources
	Reduction in use of fossil fuels and increase in renewable energy use	Reduced fossil fuel use and increased renewable energy use	Use of renewable energy	More renewable than non-renewable energy used in the project
Resource efficiency	Efficient use of resources (e.g. water, soil nutrients/fertiliser)	Reduced consumption of inputs such as water, nutrients/fertiliser while maintaining/increasing productivity	High resource use efficiency[3] e.g. water, fertiliser	Optimal consumption of inputs such as water, nutrients/fertiliser and high productivity

(continued)

Table 5.6 (continued)

Project standard	Objectives for modifying an existing project	Critical success factors for modifying an existing project	Objectives for design of a new project	Critical success factors for a new project
	Enhancement of energy efficiency	Reduced energy intensity[1]/increased energy efficiency[2]	Use of energy efficient practices	Low energy intensity and high energy efficiency in the project
	Rationalisation and reduction of transport requirements	Optimised project transport system (from input procurement to sale of produce)	Optimise transportation throughout the project value chain	Optimised project transport system (from input procurement to sale of produce)
	Reduction/minimisation of waste	Reducing spoilage and loss in the project	Minimise waste in the project	Minimal spoilage of produce and waste in the project
Social equity and inclusivity	Contribution to local food supplies	Direct food sales to local markets	Contribution to local food supplies	Direct food sales to local markets
	Contribution to poverty reduction	Provision of livelihood opportunities for the poor	Contribution to poverty reduction	Provision of livelihood opportunities for the poor
Sustainability and long term economic viability	Increase productivity and put measures in place to sustain it in the long term	Resources necessary for production such as inputs, and human resources maintained at optimum levels	Achieve and maintain high productivity in the long term	Maintaining all resources necessary for production such as water, inputs, human resources at optimum levels

(continued)

Table 5.6 (continued)

Project standard	Objectives for modifying an existing project	Critical success factors for modifying an existing project	Objectives for design of a new project	Critical success factors for a new project
	Sustained long term profitability	Sound long term management and marketing plans	Ensure profitability in the long term	Sound long term management and marketing plans
	Building adaptation to general global change and to climate change	Agricultural practices used in the project and the people who manage the project are dynamic and responsive to change	Adaptability to global and climate change	Effective climate and global change adaptation strategies and requisite capacity to implement them. Agricultural practices used in the project and the people who manage the project are dynamic and responsive to change
Relevance to the local context	Contribution to the local economy and community development	Project has role in local economy and local community development e.g. sales to local vendors	Contribution to the local economy and community development	Project has role in local economy and local community development e.g. sales to local vendors
	Contribution towards local development	Project contributes to some aspects of rural development e.g. providing jobs to rural people	Contribution towards rural development	Project contributes to some aspects of rural development e.g. providing jobs to rural people

[1]Energy intensity is a measure of the energy required per unit of output or activity (National Academy of Engineering and National Research Council 2008)
[2]Energy efficiency refers to using less energy to produce the same amount of services or useful output; or deriving more output from a given amount of energy input (Patterson 1996)
[3]Resource use efficiency can be defined as deriving maximum output per unit of resource used; or using less resources to produce a given output

rural areas, specific actions would be needed to ensure that a project contributes to rural development and local economic development. Projects could, for example, procure inputs from local suppliers, employ local people, and sell some produce to local consumers and traders, thus contributing to local livelihood opportunities. Choosing crops that are suitable for local conditions and that can be marketed with minimal losses are some of the actions to ensure a project's sustainability and financial viability. Examples of some of the actions that would be required to ensure success of green economy projects are listed in Table 5.7. In addition, the table provides potential indicators for measuring the success of these actions in terms of meeting the objectives; which will be discussed in more detail in Sect. 5.6.3.

Changing a project to meet green economy ideals is unlikely to happen instantly; instead, full alignment of a project with green economy ideals will generally only be attainable over time. However, project modification should consistently work towards improving and finally attaining full alignment.

5.6.3   Step 3: Monitoring and Evaluating a Green Economy Project

Monitoring is the ongoing process of obtaining regular feedback on the progress being made towards achieving the objectives of a project (UNDP 2009), and is based on assessing actions and progress towards achieving planned results. Evaluation, on the other hand, is an assessment of either completed or ongoing activities to determine the extent to which they are achieving stated objectives and contributing to decision making (UNDP 2009). Through monitoring and evaluation, the extent to which a project is achieving its goals can be assessed, and potential problems identified; which can in turn be used to inform decisions on actions that are required to improve performance. A green economy project would require regular monitoring to ensure that it stays on course to achieve its objectives. Specific project activities would also have to be evaluated at suitable intervals. Evaluation is critical for sustaining a green economy as it ensures that lessons from existing or completed (for example when a crop is harvested) projects are fed back into the implementation of other projects to ensure achievement of objectives.

Suitable indicators of performance are critical for the success of any monitoring and evaluation initiative. An indicator is an instrument used to describe and/or give an order of magnitude to a given condition (UNEP 2014). Indicators provide information on the historical and current state of a given system, and are particularly useful for highlighting trends that can shed light on causal relations among the elements composing a system (UNEP 2014). Both quantitative and qualitative information can be used to define an indicator, depending on the issue that needs to be analysed, as well as on the availability and quality of data (UNEP 2014). The indicators used to monitor and evaluate a project should be relevant; and should be based on the specific objectives, critical success factors and actions identified for that project. It is

Table 5.7 Illustrative actions for and indicators of success in a green economy project (as aligned to a project's objectives and its critical success factors)

Objective	Critical success factors	Actions required to achieve success	Indicators of success
Use of practices that promote soil health	Healthy soil/Improvement of soil health	• Reduce chemical fertiliser use • Increase organic fertiliser use • Reduce frequency and intensity of tillage • Control soil erosion, leaching and runoff	• Proportion of organic fertilisers used relative to inorganic fertilisers • % of total planted land area under reduced tillage • Number of practices adopted to prevent soil erosion, leaching and runoff
Use of energy efficient practices	Low energy intensity and high energy efficiency	• Reduced energy consumption of key activities such as irrigation (e.g. by using more efficient pumps and irrigating according to crop needs) • Transport inputs and produce in bulk to reduce fuel use	• Energy consumed per tonne of crop produced
Minimise negative impacts on the environment	Avoiding environmental degradation and conserving natural resources	• Avoid pollution (to soil, water and air) from agricultural practices through reduced and/or precise use of fertilisers and pesticides • Protect natural vegetation e.g. by avoiding unnecessary clearing and protection from hazards such as fires	• Number of best management practices adopted to prevent discharges of fertilisers/chemicals into water, soil and air • Fire prevention/protection measures in place • Frequency of fires
Use resources (e.g. water, fertiliser) efficiently	Reduced consumption of inputs while maintaining productivity	• Irrigate to match crop water requirements • Use drip irrigation to minimise water loss	• Amount of water used per tonne of crop produced • Quantity of fertilisers used per tonne of crop produced • Number of best management practices to improve water productivity e.g. irrigation scheduling based on crop needs adopted

(continued)

Table 5.7 (continued)

Objective	Critical success factors	Actions required to achieve success	Indicators of success
Contribute to local food supplies	Direct food sales to local markets	• Sell some of the food produced by the project directly to consumers e.g. through farm gate sales • Use local entrepreneurs to market food produced by the project	• % of produce from the project sold to markets within local community and nearest town • Proportion of produce from the project sold to local traders
Contribute to poverty reduction	Provide livelihood opportunities for local people	• Employ local people—reduce mechanisation and adopt labour intensive practices; and employ local people	• Number of people participating in the project in various capacities e.g. as employees or doing business with the project
Achieve and sustain high productivity in the long term	All resources necessary for production and human resources maintained at optimum levels	• Apply best management practices for all determinants of productivity (soil, nutrients, water, land, human resources) • Use crops and production methods that are suited to local environmental conditions	• Units of output (yield per hectare) per unit of input (such as fertiliser, seed, money • Number of and types of practices adopted by the project to increase productivity
Ensure long term profitability	Long term management and marketing plans to minimise financial risks to the project	• Diversify markets in order to reduce risks • Use procurement and marketing strategies that minimise input costs e.g. minimise transport costs by procuring inputs locally	• Number of financial and risk management systems in place for the project • Number of different markets supplied by the project
Adapt to global and climate change	Agricultural practices used in the project and the people who manage the project are dynamic and responsive to change	• Adopt practices that are appropriate for the changing environment • Build requisite capacity of human resources to meet projective objectives in the face of changing global and climatic conditions	• Number of appropriate (e.g. climate-smart) practices in place for the project • Number and diversity of skills building initiatives implemented

(continued)

Table 5.7   (continued)

Objective	Critical success factors	Actions required to achieve success	Indicators of success
Contribute to the local economy and community development	Project has role in local economy and local community development	• Procure goods and services from local suppliers • Market produce locally and help develop community based enterprises	• % of goods and services procured from local community and/or nearest town • % produce from the project sold to the local market/nearest town

important for projects to select indicators that they can easily monitor; as monitoring should ideally not be unduly burdensome. A key feature in the selection of green economy indicators is measurability. Measurability relates to the need for an indicator to reflect reality on a timely and accurate basis, and at a reasonable cost (OECD 2011b).

A selection of illustrative indicators that are relevant to the example of a vegetable production project is shown in Table 5.6. The indicators range from measurable variables, such as quantities of fertiliser applied, to qualitative factors such as the adoption of certain practices. Some indicators involve taking measurements and doing calculations, and this could be time consuming and require special instruments and skills. Ideally, the process of monitoring and evaluation should be simple but effective.

5.6.4   Step 4: Identifying and Managing Potential Risks and Challenges

A project may be unable to meet its objectives for a number of reasons, some of which may be preventable through careful planning, and others which may be completely unavoidable. It is thus important to identify factors that are likely to threaten the achievement of a project's objectives. Risks and challenges are project-specific, and should be identified within a project's particular circumstances. A project's risks need to be constantly assessed; as they are bound to change as factors such as the economy, climate, and the policy environment change. Once potential risks and challenges have been identified, a strategy for avoiding, mitigating, and/or managing each identified risk should be put in place.

An agricultural project that is being modified to meet green economy ideals my fail to meet objectives such as improvement of soil health, for example, because of lack of adequate knowledge on the part of those running the project on appropriate practices for building soil health. This risk could be addressed through capacity building and consistent information provision for those implementing green economy projects. As another example, a project's contribution to local community development may be threatened by factors such as a general lack of capacity of local people to access

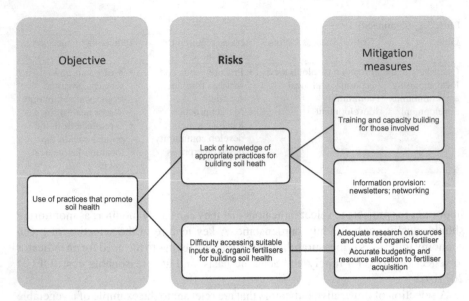

Fig. 5.3 Illustrative risks to the promotion of soil health in a green economy project and their management; and examples of mitigation measures

employment opportunities generated by a project, as a result of factors such as poverty and lack of requisite education and skills. To address such risks, a project could address capacity limitations by planning for and undertaking skills development for employees. While a project would not be expected to solve all socio-economic problems within the community, it could work with other relevant stakeholders such as Local Economic Development initiatives, local government agricultural advisory services and local businesses to build local capacity to participate in and derive benefits from local green economy projects.

For illustration, some of the potential risks to achieving the objectives of projects, and suggested means of addressing them, are presented in Figs. 5.3 and 5.4. A proactive approach to risk management, focused on preventing problems rather than waiting for them to manifest, reduces the chances of encountering problems serious enough to derail green economy project implementation.

5.7 Conclusion

Operational information that provides process level guidance on how to actually run a project minimis es doubts in green economy implementation. This chapter provides this information. A proposed framework for green economy implementation, structured as a methodical process to follow in a crop production context is described. The framework is meant for use at project level and covers factors such as setting

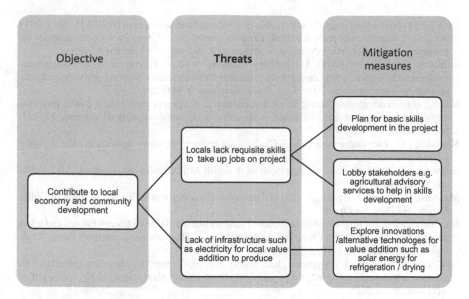

Fig. 5.4 Illustrative threats to a green economy project's capacity to contribute to local community and economic development, and examples of mitigation measures

objectives, identifying aspects that are critical for success (attaining the stated objectives), required actions, monitoring and evaluating the performance of a project; and identifying a project's risks and challenges. Options for addressing risks and challenges are discussed. The framework illustrates a green economy implementation technique that could be adapted for use in other sectors.

References

Allen C (2012) A guidebook to the green economy. Issue 2: exploring green economy principles. United nations department of economic and social affairs (UNDESA): united nations division for sustainable development

Amaruzaman S, Leimoa B, van Noordwijk M, Lusiana B (2017) Discourses on the performance gap of agriculture in a green economy: a Q-methodology study in Indonesia. Int J Biodivers Sci Ecosyst Serv Manag 13:233–247

Bellamy JA, Johnson AKL (2000) Integrated resource management: moving from rhetoric to practice in Australian agriculture. Environ Manag 25:265–280

Bellamy JA, Mcdonald GT, Syme GJ, Butterworth JE (1999) Policy review evaluating integrated resource management. Soc Nat Resour 12:337–353

Bond P (2002) Local economic development debates in South Africa. Municipal services project. Occasional papers series No 6. http://www.municipalservicesproject.org/sites/municipalservicesproject.org/files/publications/OccasionalPaper6_Bond__Local_Economic_Development_Debates_in_South_Africa_Feb2002.pdf. Accessed 16 May 2017

CGIAR (2014) Big facts: focus on food emissions. https://ccafs.cgiar.org/blog/big-facts-focus-food-emissions#.W3aTasL-vnh. Accessed 17 Aug 2018

DEA (2011) National strategy for sustainable development and action plan (NSSD 1). 2011–2014

EDD (2011). New growth path: accord 4. Green economy accord. http://www.economic.gov.za/communications/publications/green-economy-accord. Accessed 25 June 2014

European Commission (2004) Aid delivery methods: volume 1. Project cycle management guidelines. https://ec.europa.eu/europeaid/sites/devco/files/methodology-aid-delivery-methods-project-cycle-management-200403_en_2.pdf. Accessed 16 May 2017

ILO (2010) Gender mainstreaming in local economic development strategies: a guide. http://ilo.org/wcmsp5/groups/public/—ed_emp/—emp_ent/—led/documents/publication/wcms_141223.pdf. Accessed 30 Oct 2018

Kirchmann H, Thorvaldsson G (2000) Challenging targets for future agriculture. Eur J Agron 12:145–161

Koma SB (2012) Local economic development in South Africa: policy implications. Afr J Public Aff 5:125–140

Manhong ML, Ness D, Haifeng H (2011) The Green Economy and its implementation in China. Enrich Prof Publ, Singapore

National Academy of Engineering and National Research Council (2008) Energy futures and urban air pollution: challenges for China and the United States. The National Academies Press, Washington, DC. https://doi.org/10.17226/12001

NPC (2011) National development plan: vision for 2030. http://us-cdn.creamermedia.co.za/assets/articles/attachments/36224_npc_national_development_plan_vision_2030_-lo-res.pdf. Accessed 05 Jan 2015

OECD (2011a) Towards Green Growth. OECD Publishing, Paris

OECD (2011b) Towards Green Growth: Monitoring Progress: OECD Indicators, OECD Green Growth Studies, OECD Publishing, Paris. https://doi.org/10.1787/9789264111356-en. Accessed 28 Sept 2018

Patterson MG (1996) What is energy efficiency? Concepts, indicators and methodological issues. Energy Policy 24:377–390

Pretty J, Smith G, Goulding KWT, Groves SJ, Henderson I, Hine RE, King V, van Oostrum J, Pendlington DJ, Vis JK, Wlater C (2008) Multi-year assessment of Unilever's progress towards agricultural sustainability I: indicators, methodology and pilot farm results. Int J Agric Sustain 6:37–62

Richardson RB (2013) Building a green economy: perspectives from ecological economics. Michigan State University Press

Smith N, Halton A, Strachan J (2014) Transitioning to a green economy: political economy of approaches in small states. Commonwealth Secretariat, London UK

Steelman T, Nichols EG, James A, Bradford L (2015) Practicing the science of sustainability: the challenges of transdisciplinarity in a developing world context. Sustain Sci 10:581–599

UNDP (2009) Handbook on planning, monitoring and evaluating for development results. http://web.undp.org/evaluation/handbook/documents/english/pme-handbook.pdf. Accessed 1 Sept 2017

UNECA (2012) A green economy in the context of sustainable development and poverty eradication: what are the implications for Africa? http://www1.uneca.org/Portals/rio20/documents/cfssd7/1AfricaGE-BackgroundReportEN.pdf. Accessed 28 Sept 2018

UNEP (2011) Towards a green economy: pathways to sustainable development and poverty eradication. https://www.cbd.int/financial/doc/green_economyreport2011.pdf. Accessed 16 Oct 2018

UNEP (2014) Green economy: a guidance manual for green economy indicators. http://www.un-page.org/files/public/content-page/unep_indicators_ge_for_web.pdf. Accessed 13 July 2018

Vermeulen SJ, Campbell BM, Ingram JSI (2012) Climate change and food systems. Annu Rev Environ Resour 37:195–222

World Farmers' Organisation (2012) Agriculture's contribution to the green economy: proposed outcomes from the Rio +20 summit. Italy, Rome

Chapter 6
Conclusions: Key Considerations for Green Economy Project Implementation

Full Understanding and Correct Interpretation of the Green Economy Concept in the Context of Agriculture

The green economy is a fluid, theoretical concept that can be defined in several ways, and is therefore open to individual interpretation. Green economy implementation has been interpreted in various ways in the agriculture sector; with different types of practices being given the 'green economy' label, including programmes whose outcomes have been negative for the environment or for the well-being of people. Implementation should be based on understanding the green economy in terms of its fundamentals and what these mean in an agricultural context; within specific cultural, political, and socio-economic settings. Since there are many definitions of 'green economy', which could be confusing; the underlying green economy principles (see Chap. 1), should rather be used to guide implementation. Although the term 'green economy' is often interpreted in a narrow environmental sense; many of the principles of a green economy do in fact refer to socio-economic factors which are critical for human well-being. These factors should inform green economy implementation.

Context is Important

Green economy initiatives in the agriculture sector should unlock tangible social, economic and environmental benefits; and this can only be achieved if initiatives are matched to the local context and address relevant issues. The local context should be understood in terms of its social, economic and environmental components. Real life experiences of farmers are critical for understanding local contexts, as these illustrate the issues that farmers deal with on a daily basis and provide the practical context for green economy implementation. An understanding of such factors is important for crafting relevant interventions at different levels, including the farm/project and policy level. In this book, case studies of small-scale vegetable farming in South Africa have been used to illustrate local level issues that are relevant to green economy implementation in a particular context.

121
C. Musvoto et al., *Green Economy Implementation in the Agriculture Sector*,
SpringerBriefs in Agriculture, https://doi.org/10.1007/978-3-030-01809-2_6

Green economy implementation in the agriculture sector occurs within a context where global issues significantly affect local agriculture in different parts of the world. The rapidly changing and globalizing political economy of agriculture manifesting in factors such as trade in agricultural commodities in ways that are driven purely by profit motives affects commodity prices; which in turn have significant impacts on farmers (both as producers and consumers of these commodities). Furthermore, world trade in agricultural commodities is changing rapidly in terms of size, the way it is organised and the issues with which it is concerned; with global value chains increasingly playing a central role. This context presents both opportunities and constraints for farmers, who find themselves confronted with new competitive pressures, but also with opportunities to access capital, technology and markets and to enhance their skills. Global impacts can be particularly severe for small-scale farmers in Africa and other developing regions, who have little if any way of protecting themselves from what happens on world markets. Green economy implementation has to be cognisant of the prevailing global conditions and should be done in ways that enable all agricultural actors to effectively navigate global issues; operate sustainably and realise the social, economic and environmental benefits that are espoused by a green economy.

The rapidly changing technological context, characterised by the emergence of disruptive technologies driven by the Fourth Industrial Revolution has significant implications for green economy implementation. In places like Africa there are contextual challenges related to technology access and use, including the low rate of internet penetration, especially in rural areas; and inequality in access to the internet and technology on the basis of gender, location and poverty. Furthermore, the use of new technologies requires specific skills, which in many cases still need to be developed. Green economy implementation has to embrace the technological revolution, while being aware of its pros and cons and applying it in ways that ensure alignment with green economy ideals; while safeguarding the competitiveness, sustainability and profitability of agriculture.

Changing climatic conditions are already presenting challenges for the agriculture sector particularly in Africa where water shortages and yield reductions are occurring. Green economy implementation therefore has to be geared towards adapting to changing conditions; while also contributing to climate change mitigation. The adoption of appropriate practices (e.g. climate-smart agriculture) as part of green economy initiatives helps to build the capacity of agriculture to adapt to and mitigate climate change.

Green economy implementation has to be relevant to the local context, and thus has to be informed by local realities; including development issues such as creating employment and increasing agricultural output. For example, given the predominance of small-scale farming in Africa, green economy initiatives in most parts of Africa have to involve small-scale farmers. Furthermore, the actors involved in green economy implementation should fully understand issues, and not rely on generalizations. In South Africa, for example, vulnerability to poverty and unemployment varies depending on race, gender, location and educational status. Reducing unemployment in general, and for specific vulnerable groups has to be a key consideration for green economy implementation in South Africa. Within such a context, it is also important to ensure that green economy initiatives do not exacerbate existing inequalities and vulnerabilities.

An Appropriate Enabling Environment
The benefits of a green economy are only realisable through the actions of individuals and groups from multiple sectors; including government, business and civil society. An appropriate enabling environment is crucial for successful green economy implementation (realization of social, economic and environmental benefits). Flexible and responsive legislation and policies, which consider the all-encompassing nature of a green economy and linkages to different sectors, are key components of an enabling environment. In addition, to guard against reverting to entrenched practices of pursuing economic objectives at the expense of the environment and society, an enabling environment should have robust systems of accountability; as well as both incentives and disincentives for ensuring that implementation is aligned with green economy ideals.

Although policies and legislation specifically designed for green economy implementation in the agriculture sector would provide certainty and facilitate implementation, a lack of such policies does not necessarily preclude implementation. In South Africa, for example, the absence of green economy legislation has not stopped the crafting of various green economy strategies with a strong focus on agriculture (although these strategies have not yet been used in implementation). Green economy implementation can therefore draw on existing relevant policies and frameworks to provide the legislative and policy guidance required, even before sector-specific policies and legislation have been developed.

Other enabling factors for green economy implementation include security of tenure for farmers, as well as the length of the tenure. Undertaking agricultural activities linked to the green economy requires considerable long-term investment. Such investments can be risky, particularly for small-scale farmers, with limited resources. Evidence from around the world suggests that tenure security affects the willingness of farmers to invest in land and environmental protection, which has implications for the long term viability and sustainability of farming operations, and for the success of green economy initiatives. In most parts of Africa, improving women's access to land and securing their tenure rights is vital for green economy implementation, in light of the key role that women play in farming.

Socio-Economic Issues
Agriculture has a central part to play in human well-being through its direct role in food production, livelihood provision, and impacts on the environment. This makes it necessary to pay attention to socio-economic considerations when implementing green economy initiatives. Effective integration of socio-economic considerations should enable a project to achieve its direct objectives such as production on a sustained basis, and economic viability; as well as indirect socio-economic benefits, such as contributing towards poverty alleviation and job creation. Green economy projects thus have to adopt agricultural practices that not only ensure production

and profitability, but which enhance human well-being more broadly in its various dimensions, as espoused in the principles of a green economy.

Furthermore, inequalities between and within nations, and the risks and complexities associated with the impacts of development on the environment and people, places issues around governance and justice at the centre of green economy implementation. A green economy, depending on how it is implemented, could play a vital role in securing environmental justice and equality; on the other hand, it could be used as a medium for exploiting the environment and people. Green economy initiatives therefore have to be implemented in ways that safeguard individual and group rights; within the context of local social norms and values, legislation and policies; and would also have to be guided by existing safeguards, such as fair trade practices. In addressing socio-economic issues through green economy initiatives, care should be taken to maintain alignment with the full ethos of a green economy. In the case of job creation for example, the jobs created should align with green economy imperatives while also meeting local legislative requirements.

The case studies presented in Chap. 4 highlight the importance of taking local socio-economic conditions into account in green economy implementation, as these could present both opportunities and challenges. The following emerged from the case studies:

- An area with a well-developed agricultural industry would be conducive for green economy implementation, as the requisite infrastructure and support services such as input supply, transport and marketing channels would already be in place.
- An area's local economic development plans are relevant, as green economy implementation is likely to receive the requisite policy support in an area where it aligns with existing plans.
- Access to resources such as land and water for irrigation, and security of tenure, are conducive to green economy implementation. Farmers in the case studies had access to both land and water under secure tenure, and these resources did not limit agricultural production.
- The availability and state of infrastructure such as roads and electricity is critical, as it determines access to input and output markets, and access to energy required for driving machinery and equipment. The state of the infrastructure is also important as the infrastructure has to be in a usable state.
- Financial resources are required to operate green economy projects. In an area with high poverty and low income levels farmers are unlikely to have the resources to invest in green economy initiatives. Access to agricultural credit would also be difficult in such circumstances. The case study farmers had difficulties accessing credit to expand their enterprises. Green economy implementation under these conditions is likely to be hampered by the unavailability of financial resources.
- Human resources are required for green economy implementation. All the case study farmers had employees, and labour shortages were not an issue. Labour for general crop production was readily available and there is a high unemployment rate in the area.

- The case study farmers all sold their produce and had linkages to markets, and had some understanding of operating within an agricultural value chain. Such an environment would be conducive for green economy development, as the existing market linkages, capacity and experience of farmers could be leveraged for green economy projects.
- The accessibility of requisite goods and services is critical. Organic farmers in the case studies had no local suppliers of organic inputs, and had to transport them from hundreds of kilometres away from their farms (despite being within a 40 km radius of a regional town). The local availability of necessary farming inputs has implications for costs (both financial and environmental); and therefore for the profitability of an agricultural venture; as well as its carbon footprint. These issues are also relevant to green economy initiatives.
- The case study farmers were members of farmers' associations through which they accessed markets and exchanged knowledge and information. An operating environment that has institutions that are appropriate for knowledge and information dissemination would facilitate green economy implantation, as such institutions could be leveraged for the specific case of green economy projects.

The Biophysical Environment

- The biophysical environment is a key factor in green economy implementation for agricultural production initiatives, as it determines the types of crops that can be grown, and their productivity. Green economy implementation has to optimise production and use resources efficiently including selecting crops that are suited to environmental conditions. Agriculture is dependent on a healthy natural environment for the ecosystem services which underpin agricultural productivity. Green economy initiatives should be managed to ensure a healthy environment.
- There are tensions between agriculture as a primary sector, which generally has negative impacts on the environment; and the green economy imperatives related to reducing environmental risks, being low-carbon and resource efficient. Practices which sustain production while enhancing (or at least reducing negative impacts on) the resource base and environment are essential in green economy implementation. These types of practices, which include conservation agriculture, organic farming and others, are now well within the mainstream of current agricultural trends. Farmers would, however, need to be capacitated to apply these practices correctly and green economy implementation therefore has to include technical and information support systems for farmers. This support could be provided through existing agricultural technical and advisory services and farmers associations that exist in many countries.
- Green economy implementation has to balance the many benefits that humans derive from agriculture with the potential negative impacts of agriculture on the natural resource base and on ecosystems. A core part of addressing the environmental impacts of agriculture is using resources efficiently and minimising waste.

Production Methods
The case studies described in Chap. 4 highlight that agricultural production methods are a key factor in meeting green economy imperatives. Farms which used organic production methods had better alignment with the environmental principles of a green economy than farms which used conventional production methods. Green economy implementation should be backed by methods which are not damaging to the environment and align with the principles of a green economy. Farmers should, however, not be pushed to adopt a particular practice, as no single practice will necessarily achieve the objectives of a green economy in every context; rather, the emphasis should be on the fact that practices which incorporate sustainability, environmental protection and human safety are better placed to achieve green economy ideals than those that do not.

Skills, Knowledge and Information
Relevant and up-to-date knowledge and information are necessary for green economy implementation, especially in a rapidly changing global and technological context. A key aspect of green economy implementation is investment in human capital, through the provision of information, education, and skills development at all levels. This is particularly important as approaches that are compatible with the green economy may be novel to small-scale farmers. The training would also have to be tailored to specific contexts and needs. Existing agricultural initiatives with some alignment to green economy principles could also assist in sharing of lessons learned.

Realising the Potential Contribution of Agricultural Initiatives to Meeting Green Economy Principles
From the case studies described in Chap. 4, it is evident that small-scale agricultural green economy projects would have the capacity to employ people, and therefore contribute to employment creation. Employment creation contributes to the enhancement of human well-being, and is aligned with the socio-economic principles of a green economy. The case studies also highlight that small-scale agricultural green economy projects would be able to contribute to local and national economies through sales of produce and purchases of inputs. Contributions to local economies have positive impacts on livelihoods, food access and other components of human well-being; which is especially important in places with limited economic opportunities, such as rural areas. The case studies show that green economy implementation does not necessarily need to identify how to meet each individual socio-economic principle; as establishing productive sustainable agricultural enterprises which are profitable and not damaging to the environment would provide the human well-being benefits that are expected of a green economy project.

Building onto Available Resources and Infrastructure
A key lesson from the case studies in Chap. 4 is that green economy implementation would have to be based on practical realities on the ground. It also has to build upon compatible practices where these occur; and identify alternatives to practices which are not compatible with green economy principles.

Orderly Implementation Process

Relevant guiding information is vital for green economy implementation. The green economy is an emerging concept which can best be defined on the basis of a range of principles. Its implementation requires two categories of information: (i) contextual information that builds a broad understanding of what the concept entails; and (ii) operational information that provides process-level guidance on how to actually run a project. Contextual understanding enables informed application of the green economy concept to any project situation, and is vital for building capacity and flexibility in green economy implementation. Process-level information provides guidance on what to do, and is critical for minimising doubts and uncertainties.

Implementing a green economy project could be confusing, as it entails considering and translating many factors into coordinated actions. A key challenge is efficient handling the large volumes of information required for green economy implementation. A methodical process for synthesising this information and translating it into actions is critical for achieving expected green economy outcomes. The green economy project implementation framework presented in Chap. 5 of this book provides such a methodical process.

Agriculture is a well-established sector with entrenched methods and practices that have enabled the production of food and other commodities for thousands of years. These practices are not always in harmony with the environment, and green economy project implementation in the agriculture sector therefore entails making some changes to existing practices. Such changes have to be logical as well as contextually relevant if they are to be acceptable to practitioners. Modifications should not unduly disrupt production nor threaten profitability; rather, they should build on aspects that are compatible with green economy ideals, and modify those aspects that are at odds with the green economy.

A green economy initiative should be conceptualised to ensure alignment with green economy ideals from the outset. Its objectives should be clear and holistic; and cover different aspects; including issues relating to agricultural production, as well as environmental, economic and social considerations. Green economy imperatives should not be perceived as an add-on or afterthought to an agricultural initiative, but should be built into a project from conceptualization. In implementing a green economy project, it is also important to understand and define what success would look like in the specific context of the project, and to describe the indicators of success and the actions that should be taken to achieve success. Monitoring and evaluation are key aspects of green economy implementation, as they are vital for tracking progress and identifying problems. The process of monitoring and evaluation should be simple but effective, to minimise the effort required to implement projects. The costs of monitoring should not exceed the benefits.

Printed in the United States
By Bookmasters